City and Transportation Planning

Many urban and transportation problems, such as traffic congestion, traffic accidents, and environmental burdens, result from poor integration of land use and transportation. This graduate-level textbook outlines strategies for sustainably integrating land use and transportation planning, addressing the impact on land use of advanced transport like light rail transit and autonomous cars, and the emerging focus on cyber space and the role of ICT and big data in city planning.

The text also explores how we can create sustainable cities for the future. In contrast to the "compact city", which has been proposed as an environmentally friendly urban model, recent years have seen an acceleration in the introduction of ICT-based "smart city". As people's lives are drastically changed by COVID-19, a new form of city is being explored. The new concept of a "smart sharing city" is introduced as an urban model that wisely integrates physical and cyber space, and presents a way to solve future urban issues with new technologies.

Akinori Morimoto is a professor in city and transportation planning at Waseda University, Japan. He is also a vice president of the City Planning Institute of Japan and managing director at The Japan Research Center for Transport Policy. His academic field is related to the strategies to integrate the land use and transportation toward the sustainable city. He has supervised many projects in national and local government in Japan, and has been invited as a keynote speaker at several international conferences.

City and Transportation Planning

An Integrated Approach

Akinori Morimoto

Routledge
Taylor & Francis Group

LONDON AND NEW YORK

First published 2022
by Routledge
2 Park Square, Milton Park, Abingdon, Oxon OX14 4RN

and by Routledge
605 Third Avenue, New York, NY 10158

Routledge is an imprint of the Taylor & Francis Group, an informa business

© 2022 Akinori Morimoto

British Library Cataloguing-in-Publication Data
A catalogue record for this book is available from the British Library

Library of Congress Cataloging-in-Publication Data
Names: Morimoto, Akinori, author.
Title: City and transportation planning : an integrated approach / Akinori Morimoto.
Description: First Edition. | New York : Routledge, 2021. | Includes index.
Identifiers: LCCN 2021005870 (print) | LCCN 2021005871 (ebook) |
ISBN 9780367636029 (hardback) | ISBN 9780367636012 (paperback) |
ISBN 9781003119913 (ebook)
Subjects: LCSH: City planning. | Transportation—Planning. | Cities and towns—Technological innovations.
Classification: LCC HT166 .M5877 2021 (print) | LCC HT166 (ebook) |
DDC 307.1/216—dc23
LC record available at https://lccn.loc.gov/2021005870
LC ebook record available at https://lccn.loc.gov/2021005871

ISBN: 978-0-367-63602-9 (hbk)
ISBN: 978-0-367-63601-2 (pbk)
ISBN: 978-1-003-11991-3 (ebk)

Typeset in Sabon
by KnowledgeWorks Global Ltd.

Contents

Introduction   xi

1 History of cities and transportation   1

1.1   *How was the city formed? 1*
    1.1.1   What is a city? 1
    1.1.2   Who created the city? 2
    1.1.3   When was the city created? 3
    1.1.4   Where are cities formed? 4
    1.1.5   What kind of city was created? 5
    1.1.6   Why are cities created? 8
    1.1.7   How was the city created? 10
1.2   *Road and traffic history 11*
    1.2.1   Road formation and role 11
    1.2.2   Why are roads created? 12
    1.2.3   A road is formed and the city spreads 14
    1.2.4   Redesign of street space 15
References 17

2 Types of urban structure   19

2.1   *Ideal model of a modern city 19*
    2.1.1   Ideal city model 19
    2.1.2   Urban land use model 22
    2.1.3   Sustainable city model 24
2.2   *Ideal city and transportation model 26*
    *2.2.1   City and transportation model
         based on the automobile 26*
    2.2.2   The history of railway construction 30

2.2.3   Railway development model 32
2.2.4   Transit-Oriented Development 35
References 37

3 Urban structure in the next generation   41

3.1   Urban model in a declining population 41
    3.1.1   Characteristics of Japanese cities
          compared to Western cities 41
    3.1.2   Japanese city policy and compact city 42
    3.1.3   Network-type compact city 45
3.2   Hierarchy of urban structure and
    transportation system 48
    3.2.1   A transportation system that
          supports next-generation cities 48
    3.2.2   Transportation facility development
          and urban development strategy 51
References 56

4 Land use and transportation   57

4.1   Interrelationship between land
    use and transportation 57
    4.1.1   Land use and transportation 57
    4.1.2   Factors forming automobile-dependent
          cities and their countermeasures 59
4.2   Integration of land-use planning
    and transportation planning 61
    4.2.1   Relationship between land-use planning
          and transportation planning 61
    4.2.2   Population density and traffic density 63
    4.2.3   Which comes first, land use
          or transportation 65
4.3   Policy to ensure consistency between
    land use and transportation 67
    4.3.1   Land use and transportation
          integration policies in city planning 67
    4.3.2   Direct and indirect public
          intervention 68
    4.3.3   Traffic assessment 71

4.4   Toward new location guidance measures 72
    4.4.1   Location management 72
    4.4.2   Land use and transportation
          integration strategy 74
References 76

5 Consider transportation based on the city   79

5.1   What is desirable transportation? 79
    5.1.1   Transportation as derived demand 79
    5.1.2   Academic fields related to transportation 80
5.2   Transportation planning 81
    5.2.1   Traffic survey 81
    5.2.2   Traffic demand forecast 82
    5.2.3   Challenges in traffic demand forecasting 84
    5.2.4   Comprehensive transportation system 86
    5.2.5   Traffic management 88
5.3   Transportation engineering 89
    5.3.1   Understanding traffic phenomena 89
    5.3.2   Traffic simulation 91
    5.3.3   Congestion countermeasures 92
5.4   Traffic safety 94
    5.4.1   Situation and cause of traffic accident 94
    5.4.2   Analysis of traffic accidents 95
    5.4.3   Traffic safety measures 97
References 99

6 Consider cities based on the transportation   101

6.1   The essence of transportation 101
    6.1.1   Movement from the perspective
          of psychology 101
    6.1.2   Movement from the perspective
          of economics 103
    6.1.3   Transportation as a primary demand 105
6.2   Transport-based city planning 106
    6.2.1   What is the transport-based
          city planning? 106
    6.2.2   The method of transport-based
          city planning 107

6.2.3   Example of transport-based
city planning 109
6.3   Health-based city planning 110
6.3.1   City and health 110
6.3.2   Mobility in healthcare 113
6.3.3   Walking distance and utility 114
6.4   Tourism and community design 115
6.4.1   Tourism and transportation 115
6.4.2   Sustainable tourism and
community development 117
6.4.3   Methods of tourism-based city planning 119
References 121

7 Advanced transport   123

7.1   The birth of advanced transport 123
7.1.1   Thinking about new transportation 123
7.1.2   What is the advanced public transport? 125
7.1.3   What is the advanced private transport? 127
7.2   The role of advanced transport 129
7.2.1   Hierarchy of urban transportation
and advanced transport 129
7.2.2   City planning and advanced transport 130
7.3   Challenges in introducing and
promoting advanced transport 132
7.3.1   Barriers to the introduction
of advanced transport 132
7.3.2   Strategies for implementing
advanced transport 135
7.4   City planning for advanced transport 136
7.4.1   Road space for advanced
public transport 136
7.4.2   Road space for autonomous cars 137
7.4.3   A scientific approach
to implementation 140
References 143

8 Cities and logistics systems   145

8.1   Logistics planning in cities 145

8.1.1   Definition of logistics and its function 145
8.1.2   Trade and physical distribution 147
8.1.3   City and distribution channel 148
8.2   The difference between transportation
and physical distribution 150
8.2.1   Primary and derived demand of
transportation and physical distribution 150
8.2.2   Spatial and temporal movement 150
8.2.3   The difference between transportation
and physical distribution 152
8.3   City planning and physical distribution 154
8.3.1   Hierarchy of transportation
and physical distribution 154
8.3.2   Linking transportation and
physical distribution planning 155
8.3.3   Last mile transportation and delivery 158
8.3.4   Transportation and physical
distribution in the future 160
References 163
Appendix 163

9 City planning in cyberspace   165

9.1   Using ICT in city planning 165
9.1.1   Use of information and
communication technology 165
9.1.2   ICT-based transportation
and logistics planning 166
9.1.3   City planning with ICT 170
9.2   Urban models in cyberspace 173
9.2.1   Smart city 173
9.2.2   Smart cities and compact cities 175
9.3   Proposals for a new urban model that
fuses physical and cyberspaces 176
9.3.1   Hierarchy in physical and cyberspace 176
9.3.2   Smart sharing city 178
9.3.3   Building a platform to support
the city of the future 182
References 185

10 Management for future city planning   187

10.1 *Decision-making in city planning 187*
    10.1.1 *Uncertainty in city planning 187*
    10.1.2 *City planning and consensus building 188*
10.2 *Evidence-based policy making 191*
    10.2.1 *How to seek evidence in city planning 191*
    10.2.2 *City planning and city analysis 193*
    10.2.3 *City planning and artificial intelligence 196*
10.3 *City planning in moderation 197*
    10.3.1 *City planning entity 197*
    10.3.2 *The planner's philosophy 198*
    10.3.3 *Goals in the plan 201*
References 202

Index   205

Introduction

Considering the balance between transportation and land use is one of the most fundamental issues in city planning. Many of the city and transportation problems such as traffic congestion, traffic accidents, environmental impact, and lack of community are caused by poor coordination of land use and transportation. The book outlines strategies for the sustainable integration of land use and transportation, and discusses new urban models that should be pursued in the future. In particular, it touches on new technologies such as advanced transportation and information and communications technology (ICT), and proposes their impact on city planning and how to use them wisely. By using new technologies in cyberspace, I am exploring ways to integrate land use and transportation in physical space to guide cities in the desired direction. The content of this book is not only limited to generally accepted theories, but it also includes challenging descriptions of budding and new ideas. This is also because there will be situations where the conventional wisdom of city planning may not be effective in the future.

This book is a textbook for beginners studying city planning and transportation planning, as well as an instruction book for practitioners. It is also a textbook for international students to understand urban and transportation planning in more detail while learning about examples from Japan and other countries. Therefore, the emphasis is placed on acquiring basic knowledge and thinking. The study of city planning has been never complete; it constantly changes with the times. From that perspective, there are many problems that cannot be solved by existing knowledge alone. Solving new urban problems requires the ability to learn from history and established methods while at the same time generating original ideas.

Many ideas have been proposed in this book, but they should only be a hint for the reader to think about. The ability to think and propose unique solutions is crucial when faced with city planning problems with diverse backgrounds and challenges. Therefore, there is no single correct solution in city planning. I hope that even one of the ideas proposed in this book will be useful in creating a better city.

Chapter 1

History of cities and transportation

1.1 HOW WAS THE CITY FORMED?

1.1.1 What is a city?

The word "city" is often used casually, but it is difficult to find a common definition worldwide. This is also because the way of thinking of a city differs depending on the time and region. Therefore, various definitions have been proposed in literature. Some commonly agreed upon features include many people gathering in a certain area, dense buildings, and a place where secondary and tertiary industries are located. In addition, although it may be a core region regarding politics, economy, culture, and transportation, it has a low function in terms of food production. This means that a city cannot be self-sufficient and can be regarded as an area that is established only in cooperation with other areas.

According to Fujita (1993), the emergence of cities forced many agricultural villages, which had been living self-sufficiently, to change into rural areas that supplied the city with daily necessities. In other words, a city could only be established where there were rural areas that could support it. In these rural areas, surplus production became a necessity to support these newly emerging cities, and a complex political system between farmers and rulers was established. Arthur Korn (1967) also pointed out that before the emergence of full-fledged cities, there was a transitional stage such as village cities and castle cities, and that cities were formed through this transitional form throughout all periods of history.

It is said that the etymology of the word "city" comes from "Civitas". Civitas is a Latin word for a political community of citizens (cīvis) such as a "city" or a "state", or the citizenship granted by it. On the other hand, the etymology of urban is urbs, which means a set of buildings. The former is a concept of a city based on religious

Figure 1.1 Agricultural settlements, villages, conversion to cities.

Source: From Hiroo Fujita: *The Logic of the City; Why Power Needs the City*,
     Chuko Shinsho 1151, 1993 (in Japanese).

and political groups while the latter can be classified as a physical
concept of a city.

1.1.2 Who created the city?

A city is built as a result of many people gathering together. It may
have been created as a result of each individual being relatively free to
choose a place of residence, or it may have been formed intentionally
by someone for a purpose. The former is a commercial city formed
in the market, which is the result of each person's actions to maximize the individual's interest within a certain rule, with the leading
players being the citizens. The latter was constructed with political
intentions, such as the control and defense of conquered territories.
For this case, it is the rulers who determine the shape and function
of the city.

In a democratic country where the people have sovereignty,
cities are established because of the citizens' free economic activities. Elected officials and government agencies are entrusted to
work on the behalf of citizens to plan the city and develop public
infrastructure. In this case, the people who create the city are
its citizens.

On the other hand, in a monarchy governed by a specific ruler,
political intention and power structure influence the city. In some
cases, existing cities may be expanded or maintained, and in other
places, newly planned cities may suddenly appear. Cities as symbols
of authority often have a geometrical structure, which underpins the
political system of the ruler.

In any case, when planning a city, the planning theory that serves as the basis is crucial, and a city planner who has the knowledge and skills is required. The first recorded city planner was Hippodamus, who was active in Miletus around the 5th century BC. Influenced by the politics of ancient Greek philosopher Aristotle (BC 384–322), he actively adopted a grid pattern as an ideal city to preach the rational relationship between buildings and roads.

In addition, the theory of planning also has different perspectives depending on the underlying academic system. Among research on urban theory for industrial cities, urban sociologist L. Reissman (1966) divides urban experts into four types according to the two axes of problem type and data type. The problem type can be split into those who focus on solving practical problems compared to those who focus on solving more theoretical problems. The types of data can be classified into quantitative data, which can be objectively measured, and qualitative data, which is nonnumerical data. By combining these two axes, he describes four types of urban experts: Practitioners, who use quantitative data to improve specific situations; Visionaries, who use qualitative data; and Empiricists and Theorists, who explain cities theoretically and with quantitative and qualitative data, respectively, although not directly related to solving real problems.

1.1.3 When was the city created?

Lewis Mumford (1895–1990) explains in the first part of his book "The City in History" that no single definition will apply to all manifestations of the city, and that the origins of the city are obscure, a large part of its past buried or effaced beyond recovery. Due to its long history, it is difficult to accurately pinpoint the first city created. However, some experts believe that the first cities were formed around 10,000 years ago during the Paleolithic period. The transition from hunter-gatherer societies to agricultural societies allowed for the planned production of food, which increased efficiency creating a surplus of food. This allowed the first cities to form as the people living in them did not have to engage in the production of food. The timing of the formation of cities varies from region to region depending on climate, culture, industry, etc. of each region. However, according to Hibata (2008), the urbanization of the world can be divided into the following six time frames, and the names of cities are provided as examples.

1. 3000 BC–2000 BC: four major rivers (Mesopotamia, Egypt, Indus River, and Yellow River) civilization cities
   Ur, Babylon, Harappa, Mohenjo-daro
2. 8th century BC–4th century AD: Greece, Rome, China (Spring Autumn), Persian Empire
   Athens, Miletus, Priene, Olynthus, Pompeii, Pergamon, Alexandria
3. 7th–10th centuries: China (Tang), Japan (Asuka/Nara/Heian), Islamic cities
   Changan, Heijo-kyo, Heian-kyo, Fez
4. 11th–15th centuries: walled cities of medieval Europe, Japanese castle towns
   Turin, Florence, Carcassonne, Nordlingen, Palma Nova, Naarden, Edo
5. 18th century–early 20th century: Baroque city
   Paris, Vienna, Barcelona, Karlsruhe, Tokyo, New York
6. 19th century–early 20th century: modern cities of industrialized society

Looking back on the history of cities, we can see many that have risen and fallen, constantly changing as time goes on to cater to the needs of its residents. In particular, because of the increase in productivity due to technological innovation, the improvement of public infrastructure, such as roads, bridges, sanitation, and facilities, has allowed cities to grow massively in scale compared to before.

For example, before the birth of the modern city, feudalism was prevalent for a long time. In these cities, industries were centered around agriculture, forestry, and fisheries, and the city developed as a base for commercial functions based on the sales. However, as industrialization occurred, productivity increased, improving the living standards for citizens leading to the rise in democracy, which lead to an acceleration in market-based principles. In modern times, many of these metropolitan areas expanded further to form a huge cluster of cities called a megalopolis.

1.1.4 Where are cities formed?

If cities are usually born in the places where people gathered, where are cities formed? The first cities of the agricultural era were formed near fertile areas. The four major civilizations of that era were all located in downstream areas of rivers. The fertile land allowed for the production of food while the vicinity to the river and the coast

allowed for convenient trade and logistics. In order to continue sup-
plying food to the many people living in the city, various types of
water transportation such as river boats and marine ships became an
indispensable part of the city.

When war is prevalent, the ruler may elect to build a city in a
strategic location to act as a defense against invading forces. Some
cities were also formed in commercially strategic points along major
trading routes in order to encourage and take advantage of trade.
Alternatively, some cities were also formed due to religious significance or built around religious sites.

In ancient China, the position of cities and buildings was decided
based on the Feng Shui thought, which controls the flow of energy by
the position of objects. According to the Yin-Yang thought, a place
suitable for the four gods (Vermilion Bird, Blue Dragon, White Tiger,
and Black Tortoise) was considered a good terrain for building a city.
It is a good place to "store wind and collect water", with mountains
behind it, sea and rivers in the front, and hills or land surrounded
by low mountains on the left and right. For example, Chang'an in
China, and Heijokyo and Heiankyo in Japan were surrounded by
mountains on three sides, and the south was open in the city of the
four gods.

1.1.5 What kind of city was created?

Looking at the shape of the city, some cities are long and narrow,
while some are circular. In addition, there are some small cities with
a total length of several kilometers, while there are cities with lengths
exceeding several tens of kilometers from the center to the suburb.
The shape and scale of a city are defined not only by the demand for
population concentration but also by the planning philosophy for
topographical conditions and the development of urban infrastructure such as transportation.

The two main elements that make up a city are blocks and
streets. A block is formed by a collection of residential areas, and
the street connects the blocks. When a block is constructed along
topographical features, its shape is naturally irregular. A rectangu-
lar block is created only by artificial planning, and a grid pattern is
formed by continuous blocks.

It is said that the first known city to make use of the grid pattern
was Mohenjo-daro (around BC 3000). In the next section, I will out-
line how the grid-like city division was adapted to cities.

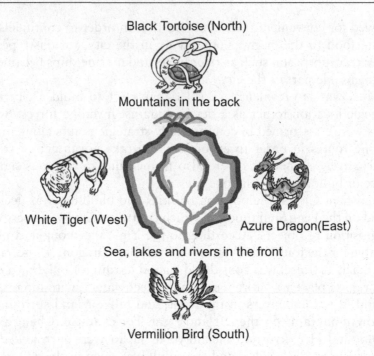

Black Tortoise (North)

Mountains in the back

White Tiger (West)

Azure Dragon(East)

Sea, lakes and rivers in the front

Vermilion Bird (South)

Figure 1.2 The city of the four gods.

Many cities of Ancient Greece were based on grid blocks. In other words, the land was first used for buildings and plazas, and roads were arranged subordinately depending on the combination. Therefore, although it was easy to expand to the outside, it had a feature where it was difficult to balance the street structure in the city.

On the other hand, the city plan of Rome was based on straight streets and the blocks were arranged accordingly. When allocating a town, first, the layout of the street and the hierarchical structure according to the width of the street were determined. The city as a whole has a good balance, but when the city expanded, there was a problem with its consistency with the outside world.

In the construction of cities in ancient China, the outer boundaries of the city were first determined, and the axis streets were set in a grid pattern. In Chang'an, the capital of the Tang dynasty (constructed in 582), the city was formed in a lattice pattern within the city walls 9.7 km east–west and 8.7 km north–south. This influence is reflected in the city of Heijokyo in Japan, built in 710, where a 36 m wide road was placed in the center with nine articles in the north and south and four articles in the east and west. Heiankyo (793), which later became the capital and now named Kyoto, was created

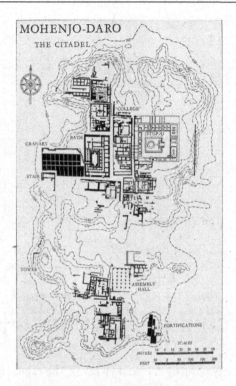

Figure 1.3 Mohenjo-daro: Plan of Mohenjo-daro citadel.

Source: Reprinted with permission from Mortimer Wheeler, *The Indus Civilization* (Cambridge University Press, 1962).

with the same idea with its city divisions based on a square (121 m × 121 m) block.

In medieval European cities (11th–13th centuries), more functional structures were found compared to geometric structures. Streets exist for walking and often have irregular arrangements for defense reasons. On the other hand, a regular grid-like street pattern is adopted for military and political planning purposes. Renaissance cities (14th–16th centuries) use a geometric structure with axes and centerlines to show the authority of the ruling class, providing a straight road suitable for the movement of artillery and troops giving a military advantage.

Afterward, as cities became more congested and the military use of gunpowder became more popular, the need for the city wall began to decrease and many cities began to spread beyond the city walls. For the city of the Baroque (17th to early 20th century), it transformed from a closed space to an open space and set a straight road with a wide road along the city axis.

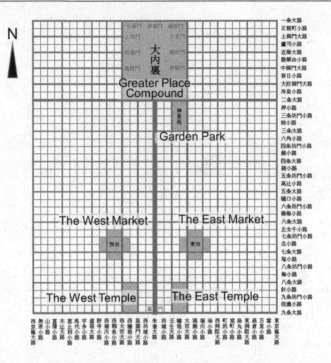

Figure 1.4 Ancient city planning in Heiankyo (793): The capital of Japan from 794 to 1869.

Source: Modified from Wikimedia Commons (https://commons.wikimedia.org/wiki/File:Heiankyo_map.png.us).

In general, whether it is a planned city with a geometrical structure based on a grid or an irregular city based on topographical features, the expansion of a city depends on the time as well as the needs of its citizens.

1.1.6 Why are cities created?

There are many reasons as to why cities are created. It seems as though the original purpose was to trade food surplus production. Food storage and the markets that buy and sell it are essential urban facilities. When the exchange of goods switched from a barter economy to a monetary economy, it became easier to trade goods, which increased the number of items traded. This in turn expanded the market and promoted population concentration. It can be said that this accumulation of various economic activities produced additional benefits, encouraging city growth even further. In urban economics, this is called agglomeration economics. When a market is created by a local population, functions such as politics, religion, education, medical care, and entertainment are added to maintain and manage the city.

Under the condition that the location of buildings can be selected freely, the process of urban structure can be explained by Hotelling's model and Alonzo's bid rent. The Hotelling model is the theory that "stores that sell products of the same price in the same trading area can ultimately make the most profit by being located close to each other". Assuming that consumers are evenly distributed in the trade area, many stores gather in the city center because the center of the trade area can acquire the most customers, and as a result, a downtown shopping area is formed. However, if many stores are concentrated in the city center, it will be overcrowded and the land price (rent) in the city center will rise. Those who cannot expect to receive income commensurate with their rent will gradually move from the city center to the suburbs. The amount of rent that can be paid out at this place is called "bid rent", and Alonzo theorized that the bid rent would decrease depending on the distance from the center of the city, and the entity with the highest price would be located. As a result, it is rational to locate land distributed in the order of retail business, industry, housing complex, detached house, and agriculture from the center of the city, and the land use distribution within the city is determined.

On the other hand, in order for a ruler to maintain the power of the government, there are cases where a city is built at a key place for trade or defense, and city facilities such as a palace are built as a

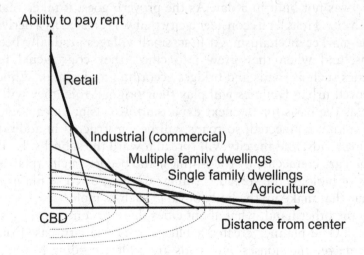

Figure 1.5 Intra-urban land use allocation in relation to bid rents.

symbol of power. In this case, a geometrical urban structure is usually preferred to symbolize the authority of the ruler. In addition, from the viewpoint of city defense, the city walls are installed to prevent the invasion of foreign enemies, and grid-like blocks are adopted for the ease of tax collection and management.

In addition, the relationship with surrounding rural areas that supply food is important for the formation of cities. Unless the city itself is self-sufficient in food, to support the urban population, a steady supply of food from surrounding areas is indispensable. While in principle it would be appropriate only for the production surplus to be supplied to cities, there were many cases where the rulers and powerful residents of cities exploited farmers when there was no food surplus. Even when the farmers suffered from food shortages, the landowner collected a certain amount of tax, and food was supplied to the urban areas. This food and goods collection system is the key to the establishment of a city, and either a political system based on the power of the governor or an agglomeration based on the market mechanism works to form a city. In other words, it can be said that the flow of power and money is the basis for creating a city.

While there are cases where money and power work together to form a city, in some cases, in order to solve urban problems such as overcrowding, an ideal city model is proposed, and cities are created systematically based on it.

1.1.7 How was the city created?

Rome was not built in a day. As the proverb goes, it takes years to build a city. First, let us consider a city that was created as a derivative of the market mechanism. At first, small villages gradually became towns, and when they grew into cities, they constructed urban facilities such as roads and bridges according to demand. While the improved urban facilities will play their original role, they will also serve as the basis for the next expansion. Roads are meaningless if they are not connected, so roads will be constructed in addition to existing roads, and the city will spread toward the suburbs. Facilities (points) are connected by roads (lines) and become a city (planes). In terms of biology, the spread is similar to the growth of filamentous hyphae that make up the body of fungi such as mold.

On the other hand, what about cities that are constructed systematically? The planner creates a blueprint of the city based on the ideal image, and houses and roads are built according to the plan. Although the shape depends on the planner's intention, it is generally constructed in consideration of the overall balance by preferring a linear shape, a grid shape, or a geometrical shape. Strong control or administrative power is required to achieve this because it evicts landowners and users and places significant restrictions on their use.

The former has a high affinity with its surroundings because the city grows while gradually changing due to the accumulation of individual economic activities, but it can take a long time to solve urban problems such as overcrowding. Since the latter develops areas in a planned manner, it approaches the ideal situation at once, but it is highly possible that it will deviate from the social needs after reaching the ideal state.

Even in the case of forming a city in a planned manner, the response to market principles gradually becomes more intense. For Baroque city planning, which started at the end of the 16th century, a mechanism for returning development profits emerged, and public institutions began to carry out projects and obtain development funds. For example, in the remodeling of Paris by French politician Haussmann (1809–1891), forcible land expropriation was implemented, while the court determined the expropriation compensation amount and a method of returning the increase in real estate prices (capital gain) was adopted. This tendency led to the beneficiary burden system (Adikes Act) introduced in 1902 in Frankfurt, and Japan will evolve into a land readjustment project independently by taking this as a model.

1.2 ROAD AND TRAFFIC HISTORY

1.2.1 Road formation and role

A road is a "place where people and vehicles pass" and is a linear space for movement. However, the Chinese characters synonymous with roads are polysemous, meaning also ways and means, lawfulness, ethics such as morality, and normativity in politics and academics. In the West, the New Testament Gospel of John 14:6 says, "I am the way and the truth and the life", indicating the path of theology. In Japanese, "Michi", which has the same character as road indicates "reasonable teachings and things that people should obey".

As humans and animals move around, grass and plants are naturally stepped on and crushed leading to a space where it is easier to travel. This is the beginning of a road. Therefore, roads have existed from the time when creatures began to walk along the ground. Roads play a crucial part in our daily lives and have many different names depending on the type. While not always a common definition, for example, a natural road is called a path, a horse-riding road was called the road, and a paved road is called a street. Not only this, but an elevated geological road is also called a slope, a narrow road is called a row, and a public road is called a highway. In Italy, the term "street" changes according to the streetscape: boulevards are called Via, tree-lined streets are called Viale, main streets are called Corso, back streets are called Vico, and streets along the shore are called Riva.

In addition, roads can be classified as private roads or public roads by the administrator, and according to their roles in the road network, they can be classified as highways, auxiliary highways, and residential roads. General roads can be used by an unspecified number of people, all of whom can enjoy the benefits unless traffic is congested. It is also difficult to exclude the use of others. In economics, such goods are called quasi-public goods.

While roads are crucial for the passage of people and goods, it also plays many other roles. Creating a road not only makes it convenient to move but it also prevents the spread of fire, secures ventilation, and can act as a playground. The various functions possessed by such roads can be roughly classified into a traffic function for movement and a spatial function for road space. Furthermore, traffic functions can be classified into traffic functions for processing traffic and access functions for entering and exiting roadside facilities in the transportation network. On the other hand, spatial functions are functions that form the skeleton of urban areas, disaster prevention functions that prevent spread of fire, living environment functions such as daylighting, and functions that coexist with urban facilities such as water and sewerage (Table 1.1).

1.2.2 Why are roads created?

The simplest reason for creating a road is the demand for traveling to different places. If a small number of people use it, it may be a simple one, but for many people and goods to move, it requires a certain width and a certain strength roadbed.

Looking back on history, the oldest road on record is the road used to transport materials for the construction of the pyramids (BC 2600). After that, a long-distance trade route (Amber Road: BC 2000) was completed from the Baltic Sea, which was an amber-producing region, to Italy, across the Alps via the Balkan Peninsula.

Table 1.1 Classification of road functions

Major categories	Subcategories	Feature description
Transportation functions	Traffic function	Ability to handle traffic flow
	Access function	Access to roadside facilities
Spatial functions	City forming function	Formation of urban structure, city block formation, and inducement of roadside development
	Disaster prevention	Evacuation routes, fire prevention
	Living environment	Daylighting, ventilation, playground
	Urban facility storage	Storage of tracks, water and sewage, power/ telephone lines, gas pipes, subways, etc.

Asphalt was first recorded to be used to build roads around BC 600 in the ancient city of Babylon. The famous Roman road (BC 312) was a state-run public road, and its length reached about 8500 km at its peak. The famous trade route centering on silk between China and the Mediterranean was called the Silk Road (BC 50). Comparing these two roads, the Silk Road was created through trade between people, while the Roman road was created systematically by high technology with a clear intention to govern the territory. Looking at the subsequent history of these two roads with contrasting reasons for their formation, the Silk Road, which was formed as a natural road, disappeared as time went by, while the Roman road, which was firmly formed as an artificial road, still retains its remains.

The amber and silk that were the targets of trade were both rare and highly valuable per unit weight, making them profitable even after deducting the cost of traveling to distant places. It is recorded that silk and gold were sometimes traded at the same weight in Rome around the 3rd century. China also brought paper and papermaking methods to the west, and trade between the east and west gave rise to an exchange of culture and technology. The reason why roads were established varies from place to place such as the transportation of goods, trade, or to rule of the occupied territories, but the road network has spread all over the world since then.

1.2.3 A road is formed and the city spreads

Economic activity will increase when the roads are open, and people and goods can easily move around. The reason for this is that the basics of the economy come from the marginal gains from the transfer of ownership. If the difference between the prices of goods in the production area and the consumption area is higher than the cost required for moving, profit is generated. Therefore, if travel costs are reduced by roads, the amount of transportation will increase, and the economy will become more active. Not only this, due to lower transportation costs, residential areas will be expanded to the suburbs where land prices are lower, and the city area will be expanded accordingly.

In this way, traffic plays a major role in shaping the city. If the main form of traffic is walking, the city can be expanded in all directions unless there are topographical restrictions. On the other hand, for mass transit systems such as railways, the expansion area is limited by the location of stations and the network. While transport supports urban activities, it has a major impact on changes in urban land use.

The size of the city depended heavily on the main modes of transportation of the era. Before the invention of motorized transportation (MT) such as railroads and cars, walking was the basic means of transportation. Horse-drawn carriages were available in some areas, but limited to the privileged classes, the affluent, or to transport goods. Two km, or around 30 minutes of walking is said to be the upper limit of distance and time that would not interfere with daily life and be an inconvenience. Therefore, the size of a city was often based around that distance. Boat transportation was also an important mode in this era, and cities were often formed by sharing the roles of road and waterways.

Steam locomotives appeared in the early 19th century, and railway construction continued throughout Europe and America. Cities and districts were connected by rail, and the shape of cities was gradually changed from a "point" to a "line" with cooperation between points. When cars became popular in the 20th century, the shape of a city, which was a one-dimensional line, expanded to a two-dimensional surface. The electric elevator, which was put into practical use in the latter half of the 19th century, encouraged the construction of skyscrapers, and the city evolved into three dimensions.

1.2.4 Redesign of street space

The term "street" is used for a road within a city, and is derived from the Latin word "strata" meaning pavement. Therefore, it has the meaning of a paved road, but here it is defined as a public space adjacent to buildings in a city and used for movement. In the United States during the motorization of the 1960s, Rudofsky was one of the first to emphasize the importance of streets in his book "Streets for People" (1969). He stated that a street does not exist where there is nothing, but only when there are buildings that frame the street. In addition, a perfect street is a harmonious space, and he discussed the relationship between people and streets based on various examples. Streets are also the blood vessels that flow through the body of the city, and can be interpreted as supporting the activities of the entire city in coordination with the capillaries that are the city's narrow streets.

Streets and buildings are inseparable, and the appropriate ratio (D/H) of street width D to building height H varies depending on the historical period classification. In medieval cities, which developed within a limited area surrounded by walls, the street-to-building ratio was roughly 0.5 because the streets were not wide enough. In the Renaissance, streets could be widened comparatively, and Leonardo da Vinci thought that $D/H=1$ was ideal. Later, in the Baroque period, the ratio of streets to buildings increased significantly to 2. In recent years, the ratio of street width to the height of buildings along the street is considered to be even at about 1–1.5, with a comfortable sense of enclosure at about 1–3, and no sense of enclosure at 4 or more.

Looking at the use of street space, the popularization of the automobile, which began in the 1920s, led to the occupation of streets around the world by automobile traffic. In cities where there was a horse-drawn carriage culture and the roadways and sidewalks were separated, walking space barely remained, but in cities where there was no division between footpaths and carriageways, the space where pedestrians could move safely gradually disappeared as the number of cars increased. Furthermore, the proportion of parking lots has gradually increased, with some cities having more than half of their urban area occupied by streets and parking lots. The number of such car-dependent cities continued to increase as people's income levels rose in many parts of the world.

On the other hand, since the mid-20th century, some cities have begun to redesign the use of their streets. This trend first appeared in European cities, where street revitalization efforts were launched to keep cars out of city centers and return streets to their original use. As a leading example, in 1962, the pedestrian street "Strøget" in the city center of Copenhagen was completed as the longest pedestrian street in Europe (1.1 km). In 1956, there were only three sidewalk cafes in New York City, but by 1968, the number had increased to 100, and signs of change in streets began to appear in the United States. Rudofsky (1969) warned against the automobile-first policy of the United States at the time and emphasized the importance of human-centered streets, saying that if Americans were to seek a dignified urban life, streets would be the first to be redesigned.

About half a century has passed since then, and large urban centers that used to be overflowing with automobiles are being reborn as human-centered streets. In 2005, the removal of highways and the revitalization of riverside areas began in Cheonggyecheon in central Seoul, and the reorganization of street space on New York's "Broadway" has been activated since 2008, showing the spread of creating recreational spaces for people in urban centers. Marunouchi,

Copenhagen Strøget (1962-)   New York Broadway (2008-)

Seoul Cheonggyecheon (2005-)   Tokyo Marunouchi (2015-)

Figure 1.6 The pedestrianization of street.

Japan's largest office district adjacent to Tokyo Station, also began pedestrianizing its streets in 2015.

REFERENCES

Hiroo Fujita. 1993. *The Logic of the City; Why Power Needs the City*, Chuko Shinsho 1151 (in Japanese).

Arthur Gallion, Simon Eisner. 1975. *The Urban Pattern, City Planning and Design*, 3rd ed., New York, NY: Litton Educational Publishing, INC.

Yasuo Hibata. 2008. *A World History of Urban Planning*, Kodansha's new library of knowledge (in Japanese).

H. Hotelling, 1929. Stability in Competition, *Economic Journal*, 39:41–57.

Arthur Korn. 1967. *History Builds the Town*, 1st ed., 1953, Percy Lund Humphries & Co., Ltd.

Sibyl Moholy-Nagy. 1968. *Matrix of Man: An Illustrated History of Urban Environment*, New York, NY: Praeger.

Lewis Mumford. 1961. *The City in History, Its Origins, Its Transformations, and Its Prospects*. New York, NY: Harcourt, Brace & World, Inc.

Leonard Reissman. 1966. *The Urban Process; Cities in Industrial Societies*, New York, NY: The Free Press.

Bernard Rudofsky. 1969. *Streets for People: A Primer for Americans*, New York, NY: Doubleday & Company, Inc.

Kenichi Takebe. 1992. *The Story of the Road I, II*, Tokyo: Gihodo Shuppan Co., Ltd. (in Japanese).

Takatoshi Tamekuni. 1999. *The History of Civil Engineering in Our Daily Lives – Cultural Directors*, Tokyo: Toyo Shoten Co., Ltd. (in Japanese).

Mamoru Taniguchi. 2014. *Introduction to City Planning; The Functions of the City and the Concept of City Planning*, Tokyo: Morikita Publishing Co., Ltd. (in Japanese).

Shun-ichi Watanabe. 1993. *The Birth of "Urban Planning" – Japan's Modern Urban Planning in International Comparison*, Tokyo: Kashiwashobo Publishing Co., Ltd. (in Japanese).

Chapter 2

Types of urban structure

2.1 IDEAL MODEL OF A MODERN CITY

2.1.1 Ideal city model

Many urban models have been proposed since the 20th century to deal with various problems facing cities. By studying various city models that have strongly influenced modern city planning, we can see the transition of urban planning into the form that is familiar to us today.

First, I would like to focus on the Garden City movement proposed by Ebenezer Howard (1850–1928) in 1898. London at the time had a poor urban environment due to the development of heavy industry and to counter this, he proposed a city model that aimed for a "marriage between urban and rural areas". To do this, he proposed a city design where people live in suburbs close to their place of work. In addition to coexistence with nature, it is highly praised for its emphasis on autonomous management of the city and community.

One of the greats who laid the foundation for modern city planning in the early 20th century was Patrick Geddes (1854–1932). After studying biology and botany, Geddes sought the importance of prior research and social research in urban planning and laid the theoretical foundation for urban research in his book "Cities in Evolution" (1915). In the first case for urban planning, he emphasized the importance of a scientific approach, and it therefore had a great influence on subsequent urban planning.

In response to the problem of urban congestion, the architect Le Corbusier (1887–1965) in "La Ville Radieuse (1930)" proposed a comfortable and safe ideal city with separate roadways and sidewalks. It created an open space around buildings while accommodating the increasing urban population due to the construction of skyscrapers. This idea of an ideal city was reflected in the Athens

Charter adopted at Congrès International d'Architecture Moderne (CIAM) in 1933, and had a strong influence on urban planning in the 20th century.

It can be said that Corbusier proposed a three-dimensional problem solution by suggesting high-rise buildings, in contrast to Howard's goal of a two-dimensional solution through Garden City. The over-crowding problem caused by the progress of urbanization continues to this day and two major directions have been shown: whether the solution will be to deal with it by spatial development toward the suburbs or by advanced utilization inside the city.

At the same time as the cities became overcrowded, the col-lapse of the local community also began to be a problem. Clarence Perry (1872–1944) theorized in "The neighborhood unit" (1929) that neighborhood relations dilute as urbanization progresses. He proposed that a neighborhood living area (around 64 ha with a radius of about 400 m) surrounded by main roads with a population of about 5,000–6,000 people should be considered a unit, consist-ing of an elementary school, a church, a community center, a park, etc. along with shops placed along the main road. This type of city planning based on the local community unit continued to be used as a basic idea of new town development.

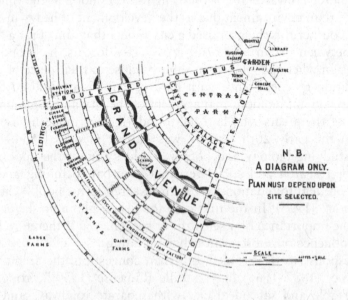

Figure 2.1 Garden City.

Source: Ebenezer Howard. *To-morrow* (Routledge/Thoemmes Press, 1898).

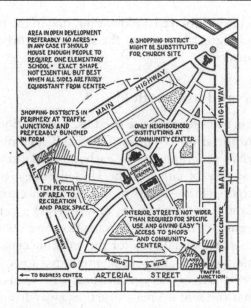

Figure 2.2 The neighborhood unit.

Source: C.A. Perry. *Neighborhood and Community Planning* (Reginal Plan of New York and its Environs, 1929).

In the 1950s, criticism of modern city planning began to arise. American-born Canadian Urbanist and writer Jane Jacobs (1916–2006) criticized the US urban redevelopment policy by describing it as a giant chess game that was destroying the community. She pointed out the importance of diversity in big cities and claimed that four conditions were needed: diversity areas, small blocks, old buildings, and concentration. Her book, The Death and Life of Great American Cities (Jacobs 1961), was received greatly as it had a unique idea of placing emphasis on the diversity of cities, as opposed to the uniformity and efficiency. This book had a large impact on the ideas of urban planning in the latter half of the 20th century.

At the same time, Kevin Lynch (1918–1984) presented a method of visualizing the city in "The Image of the City" (1960). Based on a fact-finding survey, he named the five components of the image of the city: paths, edges, districts, nodes, and landmarks. He proposed that it was desirable for cities to have a public image that is easy to understand (legibility), easy to recall (image-ability), and shared with people, in contrast to the planning theory that presented the physical way of cities. This idea played a pioneering role in the study of urban images and had a great impact on the subsequent urban design and landscape engineering.

2.1.2 Urban land use model

There are certain rules for land use in cities. The relationship between cities and transportation is outlined through an important model proposed in urban geography in the early 20th century.

Based on a survey of cities in the United States, urban sociologist Ernest Burgess (1886–1966) proposed a concentric ring model in 1925. For land use distribution, he proposed a central business district in the center of the city with transition areas with mixed uses around it. With a residential area for the working class, a residential area for the middle class, and a residential area for commuters further toward the suburbs. This is consistent with the bid rent theory, which states that the rent decreases further away from the center of the city toward the suburbs. This model is unique in that the land use distribution depends only on the distance from the city center.

Later, land economist Homer Hoyt (1895–1984) proposed a sector model in 1939 with modifications to the concentric model. This model expressed that the city grows not only in the distance from the center of the city but also in a fan shape along the main transportation routes such as railways and roads. This characteristic can be seen in cities that have developed due to traffic restrictions, and in

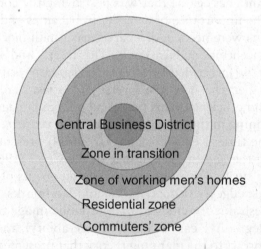

Central Business District

Zone in transition

Zone of working men's homes

Residential zone

Commuters' zone

Figure 2.3 Concentric zone model (Burgess model).

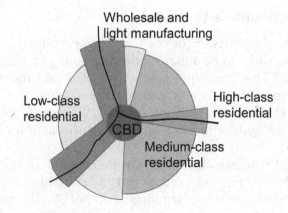

Figure 2.4 Sector model (Hoyt model).

many cases apply to British cities. It can be understood as a model mainly for railways.

As cities further develop, land use distribution will take on a more complex form. Urban geographers Chauncy Harris (1914–2003) and Edward Ullman (1912–1976) introduced the multiple nuclei model in 1945. This was a model that showed the center of a city as not consisting of one core, but that of multiple cores such as commerce and business. This reflects the time where the rapid spread of automobiles drastically reduced the spatial restrictions on land use, resulting in aggregation and dispersion of the functions of the city.

The transition of the urban land use model shows that the change of the main transportation system from the railroad to the car in the early 20th century had a great influence on the urban structure itself.

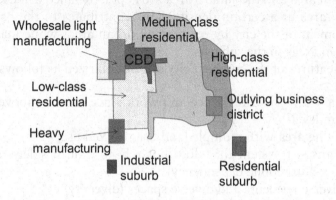

Figure 2.5 Harris and Ullman's multiple nuclei model.

2.1.3 Sustainable city model

In the 1970s, the limits of conventional mass production and mass consumption came to be a major talking point. The Club of Rome announced, "The Limit to Growth" in 1972 and stated that "If population growth and environmental trends continue as before, global growth will be reached within 100 years". This report received worldwide attention and began discussions on global sustainability.

The World Commission on Environment and Development at the United Nations issued a report in 1987 entitled "Our Common Future". It was named the Brundtland Report by the name of the chairman and the Norwegian Prime Minister. Within the report, the importance of sustainable development was identified and defined as development that meets the needs of the present without compromising the ability of future generations to meet their own needs. Since then, the concept of "sustainable development" has become widely known around the world, and the European Commission has recommended a compact urban form for sustainable development in the Green Paper on the Urban Environment (Commission of the European Communities 1990).

In addition to conventional issues of urban planning such as overcrowding, environmental improvement, and community conservation, sustainability for the future has become an important goal. Especially since the 1990s, the "compact city" concept has been attracting attention as a sustainable urban model for reducing environmental impacts with policy toward its realization being led mainly by developed countries. In a compact city, the city area does not randomly expand toward the suburbs in a disorderly manner, but the urban area and the suburb area are kept in balance while keeping the city area at a certain density. This contributes to the balanced development of the city by revitalizing urban areas while maintaining green areas in the suburbs.

The features of the compact city are summarized as follows.

1. High density of residence and work place (density above a certain level)
2. Living area with multiple land uses (mixed use)
3. Transport system that does not rely too much solely on cars (low-automobile dependency)
4. Diverse residents and diverse spaces (diversity)
5. Unique regional space (history and culture)

In the latter half of the 1980s, "New Urbanism" was proposed in the United States due to criticism of automobile-dependent suburban housing development. A similar concept is the "urban village" concept that began in the early 1990s in the United Kingdom, with a priority on public transportation and assuming living close to work and other necessary amenities. These new concepts are all aimed at returning to traditional urban planning centered around railway stations.

While the compact city was proposed as a sustainable urban model, it has been criticized for its poor feasibility and the unclear shape and conditions of the city. In addition, the urban form does not require a clear urban boundary between urban and suburban areas like in medieval cities, and its urban core can be either monocentric or polycentric. An alternative form of a sustainable city according to Haughton and Hunter (1994) has been proposed as seen in the Figure 2.6. A sustainable urban structure is a complex combination of various factors, such as natural topography, urban formation process, and resource distribution, and must be proposed with due consideration of local conditions.

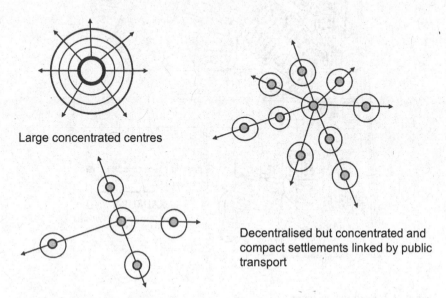

Large concentrated centres

Decentralised but concentrated and compact settlements linked by public transport

Dispersal in self-sufficient communities

Figure 2.6 Alternative form of sustainable cities.

2.2 IDEAL CITY AND TRANSPORTATION MODEL

2.2.1 City and transportation model based on the automobile

It is no exaggeration to say that the 20th century was the century of automobiles in transportation. When the Model T Ford was launched in 1908, it proved the effectiveness of mass production in factories as the price of cars fell drastically, so much so that the automobile went from being a symbol of wealth to becoming a common site to see on roads by the 1920s in the United States (known as motorization). At the same time, traffic congestion and traffic accidents

Figure 2.7 Radburn system.

Source: Radburn N. J. Plan of Northwest and Southwest Residential District (November 1929).

caused by automobiles began to emerge as social problems. The first proposal for solving this problem with urban planning was the concept of "the neighborhood unit" by Clarence Perry (1929). He suggested that it was possible to create a safe living environment by not allowing traffic to pass through neighborhoods and to provide the necessary amenities within the walking range of residential areas. In 1928, the idea of a neighborhood unit was practiced in Radburn, NJ, United States, and many dead-end roads (cul-de-sac) were adopted in residential areas. This thorough pedestrian separation system was called the Radburn system. The design of pedestrian separation at the block level within the neighborhood unit had a great impact on the subsequent housing development.

As motorization spread across the globe, responding to automobile traffic became an essential issue in urban planning. Particularly in Europe, automobiles overflowed the existing urban areas, and dealing with them became an urgent issue. In Germany in the 1950s, a zone system that eliminates passing traffic was introduced as a method of solving traffic problems while utilizing conventional streets. In addition, the British government also studied how cities and transportation should interact and summarized their findings in 1963 as "Traffic in towns". The space for passing traffic (corridor in the city)

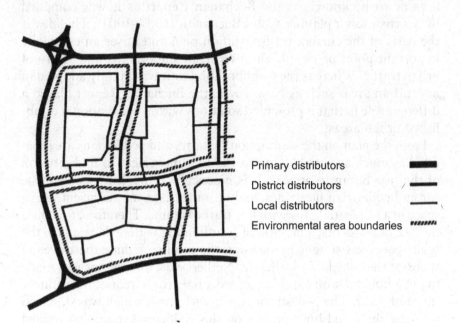

Primary distributors

District distributors

Local distributors

Environmental area boundaries

Figure 2.8 The hierarchy of distributors in Buchanan report.

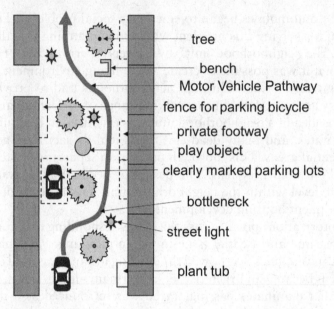

Figure 2.9 Image of Woonerf.

and good living space (room in the city) are clearly classified, and the importance of the hierarchy of the road is explained. This report became to be known as the Buchanan Report as it was compiled by Scottish town planner Colin Buchanan (1907–2001). The idea is the basis of the current transportation plan and raises an especially important point of view both in terms of smoothness and safety of urban traffic. Whereas the neighborhood unit was mainly adopted in new urban areas such as New Town, the Buchanan Report played a different role in that it presented solutions within conventional established urban areas.

From the plan on the assumption that cars and pedestrians are separated as much as possible, in the 1970s, a movement to find a fusion of the two began to appear. It is practically difficult to separate the car and pedestrian flow lines, and in some cases, an inefficient use of space in a residential area with low traffic volume. Therefore, to reduce the risk of accidents, a method of coordinating with pedestrians in the road space was sought by devising the road to reduce the traveling speed of the vehicle. In Delft, the Netherlands, an attempt was made to put a flowerpot on a road in a residential area to reduce the running speed of a car. The pedestrian-coexistence road, which was designed to lower the speed limit of cars on this residential road, was named

Figure 2.10 Shared space (Seven Dials in London, United Kingdom).

"Woonerf", which means "garden of living" in Dutch. In 1976, the Dutch Road Act and Road Transport Act were amended and Woonerf was legally recognized and the idea spread throughout Europe.

Controlling the speed of automobiles on community roads is called "Traffic Calming" and is implemented all over the world. Especially in the case of speed control in residential areas, it is called Zone 30 to indicate the upper speed limit, and physical speed control devices such as humps that raised a part of the road was introduced in the area.

In the 1980s, traffic control such as signs, signals, and the medians, which were considered essential for the smooth running of cars, came under scrutiny. Dutch transportation expert Hans Monderman (1945–2008) advocated the importance of creating a safe transportation space through pedestrian-vehicle communication and proposed the concept of a "shared space". The traffic signals and pedestrian crossings were removed from the road to call attention to users, which will contribute to traffic safety. This method is practiced not only in the Netherlands but also in the United Kingdom and elsewhere and contributes to the reduction of traffic accidents.

In general, in the 20th century, the sovereignty of road use gradually shifted from pedestrians to cars due to the explosive spread of cars. At first, many urban planners struggled with the spatial separation of the two, and various countermeasures were created. However, in the latter half of the 20th century, the integration of the two began to be gradually promoted, starting with the residential roads, and the sovereignty of road use is returning to pedestrians in the 21st century.

2.2.2 The history of railway construction

Railways are another transportation mode, which has been prominent since the 20th century. Railroads were put into practical use about a century earlier than cars, but there was a vast difference in the scale of its implementation depending on the time and region.

The first person to drive a steam locomotive was said to be Richard Trevithick (1771–1833) in England in 1804. The world's first commercial operation of steam locomotives (total length of 40 km) began in 1825 and a railway network connecting major cities was rapidly formed in the 1840s. In the early days of the railroad, many small businesses were prominent, but mergers continued with subsequent acquisitions of competitors and most existing railway companies were consolidated into four companies by the Railway Law in 1921. On the other hand, when cars started to become more prominent around the 1920s, railroad revenue decreased, and the industry gradually deteriorated. It became difficult to continue as a profitable business, and in 1948 after the war, four companies were nationalized and became the British Railways under the umbrella of the British Transport Commission. In the 1960s, the management situation worsened, and the line was forced to be abolished, following the privatization of British Railways by the 1993 Railway Law.

The first urban public transportation in the United States had appeared in 1827, when regular stagecoaches began operating up and down Broadway in New York. Initially, railroads around the United States pulled their trains with horses, but steam locomotives began operating in 1830. Compared with European cities, the United States had much more undeveloped land with infrastructure such as roads and canals rapidly being developed as a means of transporting people and goods. The construction of steam locomotives as one of the few means of transportation became a business that made a lot of money and a railway construction rush was

created. In particular, in the east of the Mississippi River in the
United States, a railway network was developed, and a route network was completed by the 1850s. The eastern part of the United
States was dense and the profitability of the railway business was
high, but the profitability in the western part was low, and the
government agency took measures through investment and free
provision of national land. Railway companies received these benefits and were able to increase profitability by selling land that was
given to them. Such a system continues to this day with a mechanism for returning development profits. As a result, it contributed
to the increase in population along the railway, the development
of industry, and the development of the railway network that provides transportation and trade across the vast land. However, in
the 1920s, the railway network began to gradually shrink, just as
in the United Kingdom. The construction of interstate highways,
the operation of highway buses, and the introduction of aircraft
in the 1950s accelerated the decline of railway users, and due to
the simplification of the abolition procedure by the Transport Act
of 1958, the number of railroad services decreased and railroads
were abolished. In the 1970s, railway companies continued to fail,
which led to the federal government nationalizing unprofitable
passenger transportation to create Amtrak. Not only this, freight
transportation, which originally accounted for the majority of
rail transportation in the United States, began revitalization due
to the deregulation of railways in the 1980s.

In Japan, the railway was first opened between Shimbashi station
and Yokohama station in 1872 with technical support from the
United Kingdom. In 1881, a semi-governmental Japanese railway
was established, which led to similar private railways being created
throughout Japan. The Railway Nationalization Law was promul-
gated in 1906 due to the difficulty of managing of private railways
and the necessity of railway management as a national policy, and
nationalization was implemented for major domestic trunk lines (17
major private railway companies). In the 1930s, the domestic railway
network was enriched with the increase in the number of limited
express trains, and the major railway lines of major private railways
in big cities were largely completed. After World War II, Japan entered
a period of high economic growth in the 1950s, and the demand for
railways increased significantly along with the increase in economic
activity. With the opening of the high-speed railway (bullet train) in
the year of the 1964 Tokyo Olympics, Japan entered the era of inter-
city high-speed railway ahead of the world. On the other hand, the

freight business, which was a major source of income for the national railways, became unprofitable due to the adverse effects of motorization, and deteriorated the management status. In 1987, Japanese National Railways (JNR) was privatized by being divided into six regions.

Railroads, which were popularized in the 19th century, were initially superior to conventional transportation, which allowed high business profitability. Many private companies engaged in the railway business, repeatedly forming mergers and integrations, and forming networks throughout the country. However, when new transportation modes such as cars and airplanes were established, the demand for railways fell relatively and the business situation continued to deteriorate. To protect the industry, many countries attempted nationalization, but ultimately resorted to revitalization efforts through privatization due to business efficiency. This is exactly what happened in some developed countries, including Japan, which was able to build a railway network like no other country in the world.

2.2.3 Railway development model

Looking back on the history of railways over the past two centuries, the railway business has certainly experienced both prosperity and decline, but it can be analyzed differently from the perspective of city planning. It is a fact that the railway network contributed to the development of the national land by providing a link between cities that had been divided up until the 19th century. In addition, the newly constructed station promoted the conversion of the surrounding land to residential land and played a role similar to a saucer for the increasing urban population. The proverb "Every miller draws water to his own mill" was imitated to create the saying "Every politician draws railway to his own region", which was used during the Meiji period in Japan. It shows that the political bargaining was actively carried out because railway construction brings great benefits to the local economy.

The history of railway construction and urban development will be outlined based on the case of Japan. In Japan, it was not the national railways, but the private railways in big cities that actively developed towns along railway lines. This is because the national railroads had a major role in transporting the main lines that traveled all over Japan and were reluctant to transport commuters in the city. The main line was a public railway, and the local transportation was a private railway. Initially, JNR's business was limited to the railway and related transportation business by the Japanese National Railways Act, and therefore it was not allowed to take part in urban development.

It was Kozo Kobayashi (1873–1957), the president of the Minoh Arima Electric Railway (later Hankyu), who built the first business model in Japan that integrated the railway and urban development in its business. Kobayashi built around 11 ha of land around the suburb station (Ikeda Muromachi station) about 20 km from the center of Osaka before laying the railway and started the sale of residential land in 1910 immediately after the railway opened. Contrary to the city center where the environment had deteriorated, the proposal of a healthy lifestyle by living in the suburbs met the needs of the times and the sale of residential land was a great success. The railway development model that started in Osaka later promoted the development of major private railways in the suburbs and contributed greatly to the construction of railway networks by the private sector.

Eiichi Shibusawa (1840–1931), on the other hand, created the first planned city along a railway line in Tokyo. Shibusawa learned Howard's theory of Garden City by visiting Europe and established Garden City Co., Ltd. (later Tokyu) in 1918. They first purchased around 1.32 million square meters of land in the suburbs of Tokyo (such as Denen-chofu and Senzoku) at a price about 600 times less than the land price in central Tokyo at that time. In addition, a subsidiary (Meguro Kamata Electric Railway Co., Ltd.) was set up to provide transportation access to central Tokyo. Keita Goto (1882–1959), who was entrusted with its management, began construction of the railway and the first line was opened in 1923. After that, the railway construction was further promoted based on the profits from the sale of land along the line.

The characteristic of the railway development model is not just the development of land around the station, but a business model that assumes the efficiency of the entire railway network. For the target line connecting the city center and the suburbs, building large commercial facilities (department stores, etc.) at the terminal station in the city center while also developing houses in the suburbs around the stations was crucial. This created not only weekday commuting traffic but also holiday shopping traffic. In addition, in order to correct the bias of users (concentration of demand only in the city center direction in the morning), which is a common issue in the railway business, bidirectional demand

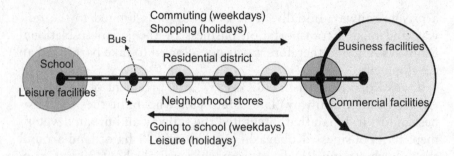

Figure 2.11 Concept of railway development model.

was created by attracting industrial parks, universities, and leisure facilities in the suburbs. It also fostered neighborhood stores
around the station and expanded the hinterland of the residential
area through bus routes, establishing a lifestyle of living along
the railway line. With the strengthening of the management base
through such cooperation between the railway business and the
development business, nonrailway revenues for private railway
companies accounted for 30% to 50% of total revenue.

The National Railway implemented a drastic plan in response
to the increase in commuting traffic in the Tokyo metropolitan area in 1964 known as the "Tokyo Five-sided Operation",
which aimed to develop station buildings using legal amendment. However, the target was mainly within the railway facilities, and even after privatization in 1987, the scope was limited.
An enlarged map of the urban area of Tokyo from the 1920s to
the 1950s, when development along the railway lines began in

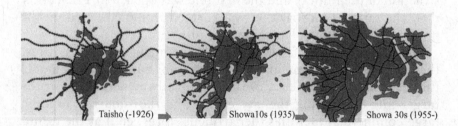

Figure 2.12 Expansion of urban areas in the Tokyo metropolitan area. Historical
review of urban growth.

Source: Modified from Ministry of Land, Infrastructure and Transport (Urban
Transport Facilities in Japan, 2002).

Tokyo, shows how the urban area of Tokyo spread to the suburbs as the railway network expanded.

2.2.4 Transit-Oriented Development

Originally developed in Japan at the beginning of the 20th century, development along the railway line attracted attention as TOD (Transit-Oriented Development) began to gain popularity in the latter half of the 20th century. New Urbanism pioneer Peter Calthorpe (1949–) proposed the concept of TOD in his book "The Next American Metropolis" in 1993. According to his book, TOD is described as follows.

A Transit-Oriented Development (TOD) is a mixed-use community within an average 2,000-foot walking distance of a transit stop and core commercial area. TODs mix residential, retail, office, open space, and public uses in a walkable environment, making it convenient for residents and employees to travel by transit, bicycle, foot, or car.

Summarized in this book, the principles of TOD are to:

- organize growth on a regional level to be compact and transit-supportive;
- place commercial, housing, jobs, parks, and civic uses within walking distance of transit stops;
- create pedestrian-friendly street networks which directly connect local destinations;
- provide a mix of housing types, densities, and costs;
- preserve sensitive habitat, riparian zones, and high-quality open space;
- make public spaces the focus of building orientation and neighborhood activity; and
- encourage infill and redevelopment along transit corridors within existing neighborhoods.

Later, focusing on the 3Ds (density, diversity, and design) as characteristics of TOD, the relationship with travel demand was investigated, and it was found that there is a certain relationship between the built environment and travel demand. These studies confirmed that creating compact, diverse, and pedestrian-oriented neighborhoods can be effective in changing the over-reliance on automobiles. TOD is a reflection on the excesses of our car-based society and a trend toward urban regeneration, and the concept

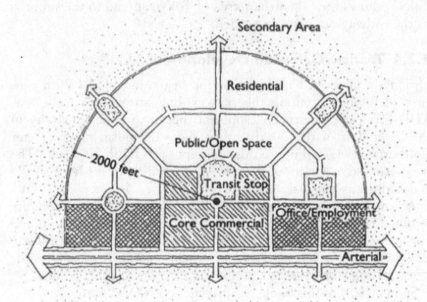

Figure 2.13 Conceptual diagram of Transit-Oriented Development (TOD).

Source: Reprinted with permission from Peter Calthorpe. *The Next American Metropolis-Ecology, Community, and the American Dream* (Princeton Architectural Press, 1993).

has spread to other parts of the world, giving rise to a variety of development examples, accompanied by the regeneration of trams and other public transport systems.

Whereas Japan's railway development along the railway is a development model essentially established before motorization, TOD is a model aimed at rehabilitating public transportation and revitalizing communities centered on walking in a car-dependent society. It should be noted that the background and purpose are quite different. In particular, Japan's railway development in the early 20th century was development at the time when there was almost no influence of cars, and the population growth due to urbanization was the basis for growth. On the other hand, it can be said that TOD in the latter half of the 20th century was based on the need to change from an excessively car-dependent society into a more sustainable society. Taniguchi (2018) states that unlike railway area developments in Japan, in which private-sector

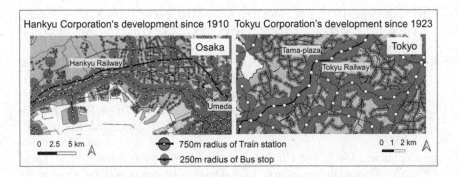

Figure 2.14 Current public transportation network in the railway area developments.

railway operators develop areas along their railways as their own domains, TOD is based on individual contracts signed between government-run railway operators and developers of areas around train stations.

Moreover, while Japan's railway development was mainly aimed at sustainability as a railway business, TOD was aimed at new urban development projects and community formation around transit. In particular, the development benefits of the former are mainly attributed to railway companies, and the latter is widely returned to society. For this reason, it is difficult for the latter railway business to be self-sustaining, and the costs for construction and operation are often widely covered by taxes, which makes social consensus extremely important. The figure shows a part of the current railway and bus network of Hankyu Corporation (1910–) and Tokyu Corporation (1923–), the first companies in Japan to implement railway development. Over the course of a century, rail and bus routes have expanded to cover the entire urban area, and the interconnected and extensive public transportation network has become an integral part of people's daily lives.

REFERENCES

Emest W. Burgess. 1925. The Growth of the City: An Introduction to a Research Project. In *The City*, eds. Park, R. E., Burgess, E. W., 47–62. Chicago, IL: University of Chicago Press.

Peter Calthorpe. 1993. *The Next American Metropolis: Ecology, Community, and the American Dream*, New York, NY: Princeton Architectural Press.

Robert Cervero, Kara Kockelman. 1997. Travel Demand and the 3Ds: Density, Diversity, and Design, *Transportation Research Part D: Transport and Environment*, 2(3):199–219.

Commission of the European Communities. 1990. *Green Paper on the Urban Environment*. http://aei.pitt.edu/1205/ (accessed August 10, 2020).

Le Corbusier. 1987. *The City of Tomorrow and Its Planning*, New York, NY: Dover Publications, Inc. (Originally published by Payson & Clarke Ltd, New York, n.d., 1929).

Robert T. Dunphy, Robert Cervero, Frederic C. Dock, Maureen McAvey, Douglas R. Porter, Carol J. Swenson. 2004. *Developing Around Transit: Strategies and Solutions That Work*, Washington, DC: Urban Land Institute.

Hildebrand Frey. 2001. *Designing the City: Towards a More Sustainable Urban Form*, London: Spon Press.

Patrick Geddes. 1915. *Cities in Evolution: An Introduction to the Town Planning Movement and to the Study of Civics*, London: Williams & Norgate.

Peter Hall. 1982. *Urban & Regional Planning*, 2nd ed., London: Unwin Hyman Ltd.

Chauncy D. Harris, Edward L. Ullman. 1945. The Nature of Cities, *Annals of the American Academy of Political and Social Science*, 242:7–17.

Graham Haughton, Colin Hunter. 1994. *Sustainable Cities*, London: Jessica Kingsley Publishers.

Ebenezer Howard. 1898. *Tomorrow*, New York, NY: Routledge/Thoemmes Press.

Homer Hoyt. 1939. *The Structure and Growth of Residential Neighbourhoods in American Cities*, Washington, DC: US Federal Housing Administration.

Jane Jacobs. 1961. *The Death and Life of Great American Cities*, New York, NY: Random House.

Kiyonobu Kaido. 2001. *Compact City: Towards a Sustainable City*, Tokyo: Gakugei Shuppan-Sha (in Japanese).

Mitsuhiko Kawakami. 2012. *City Planning*, 2nd ed., Tokyo: Morikita Publishing Co., Ltd. (in Japanese).

Hisashi Kubota, Takashi Ohguchi, Katsumi Takahashi. 2010. *Read and Learn Traffic Engineering and Planning*, Tokyo: Rikoh Tosho Co., Ltd. (in Japanese).

Kevin Lynch. 1960. *The Image of the City*, Cambridge, MA: The MIT Press.

Donella H. Meadows, Dennis L. Meadows, Jørgen Randers, William W. Behrens III. 1972. *The Limit to Growth*, A Report for the Club of Rome's Project on the Predicament of Mankind, New York, NY: Universe Books.

William D. Middleton. 2003. *Metropolitan Railways: Rapid Transit in America*, Bloomington, IN: Indiana University Press.

Ministry of Land, Infrastructure and Transport. 2002. *Urban Transport Facilities in Japan.*

C.A. Perry. 1929. *The Neighborhood Unit: Neighborhood and Community Planning, from the Regional Plan of New York and its Environs*, New York: Routledge.

David Rhind, Ray Hudson. 1980. *Land Use*, London: Methuen & Co. Ltd.

Clarence S. Stein. 1957. *Toward New Towns for America*, Cambridge, MA: The MIT Press.

Yoshiharu Takamatsu. 2015. *This Is How the Railroad Network in Tokyo Was Created: The Five-Pronged Strategy That Changed Tokyo into Greater Tokyo*, Tokyo: Kotsu Shimbunsha Shinsho, 080, Transportation News Co., Ltd. (in Japanese).

Mamoru Taniguchi. 2018. Ensen Kaihatsu (Railway Area Developments) in Japan: A Comparison With Transit-Oriented Development (TOD). In *Routledge Handbook of Transport Asia*, eds. Junyi Zhang, Cheng-Min Feng, 287–295. London and New York: Routledge.

World Commission on Environment and Development. 1987. *Our Common Future*, London: Oxford University Press.

Takashi Yajima, Hitoshi Ieda (eds.). 2014. *The World City Created by Railways, Tokyo*, Tokyo: The Institute of Behavioral Sciences (in Japanese).

Chapter 3

Urban structure in the next generation

3.1 URBAN MODEL IN A DECLINING POPULATION

3.1.1 Characteristics of Japanese cities compared to Western cities

There is a historical approach and a comparative approach to grasp the characteristics of city planning. The historical approach is a way to capture changes in a country's planning philosophy and technology in relation to its social background. The comparative approach is a method of comparing urban plans of multiple countries and identifying characteristics from problems and issues in technology transfer to other countries.

First, the characteristics of Japanese and Western cities are compared based on the cultural background. Climate (natural environment) and other natural factors play an important role in determining the location and distribution of the city, the character of society, and even its rise and fall, which can indirectly affect the structure of a city. Focusing on the differences in the processes by which humans naturally work, overcome, and create cities, many cities in the West have been formed to overcome the harsh natural environment by human intelligence, whereas in Japan, many cities have merged with nature. It can be interpreted as the difference between the western world, which created a space for human activities to counter the severe natural environment, and Japan, which gently formed a society in a warm natural environment. Barrie Shelton (2012), who conducted a comparative study of the West and Japan, said, "Western people, through their buildings and cities, conquered climate, landscape and ultimately time with acts of strength and subjection. Further, they have struggled to maintain that domination through repair and improvement. Meanwhile, the Japanese through their buildings were collaborators with all three: they touched lightly on the land, admitted the elements and succumbed to time". The difference can be seen in the way water is handled in landscaping. While there are many geometrical ponds and fountains in Western gardens, traditional Japanese gardens are shaped around a pond that takes advantage of the undulations of the land, expressing the way water flows from above. In addition, the building materials were mainly made of wood in Japan, as opposed to Western brick, and it was premised on rebuilding within a certain period. It can also be said that the frequency of occurrence of natural disasters such as earthquakes and typhoons was oriented toward structures that are more flexible and resilient than solid buildings.

Looking at the city from a bird's eye view, we notice that the urban areas often suggest geometric designs based on a kind of artificial pattern. On the other hand, from the point of view of each resident, individual logic is mixed, and the prosperity rises and falls over time, and changes beyond a certain shape. If the former influence is strong, the shape of the urban space is maintained, and if the latter influence is strong, the urban space changes.

In contrast to the clear separation of cities and suburbs in the West, Japanese cities do not have a clear distinction between the center and the periphery, with an intermediate region extending between them. It is a different evaluation axis, whether to see the unified landscape of the West as beautiful or to be attracted to the cityscape where the back alleys of Japan overlap. Another point of view in city planning is whether it is important to pass on the traditional culture or to integrate and change diverse cultures. Of course, there is no single answer in city planning, and the model answer may be "keep what we should protect and change what we should change".

3.1.2 Japanese city policy and compact city

Japan is an island country surrounded by the sea, with forests occupying about two-thirds of the country's land, and cities have developed mainly on the limited flat land between the mountains and the sea. Therefore, the city area originally had a compact shape, but when the population rapidly increased in the 20th century, the green forests and the sea were reclaimed to form a new city area. At the same time, the influx of population from the rural areas to the urban areas continued and bloating and overcrowding of the cities became a major problem.

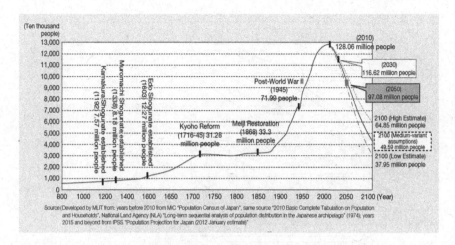

Figure 3.1 Long-term changes in the Japanese population.

Source: White Paper on Land, Infrastructure, Transport and Tourism in Japan
(2012, p.1).

The basis of land use policy during the period of population increase
was to separate the urbanized area that performs urban activities
from the urbanized control area that suppresses it. The basic frame
of land use planning in Japan is to divide land use into three levels:
City Planning Area, Urbanization Areas, and Land Use Zones. First,
a city planning area, which should be improved, developed, and pre-
served, is defined as a unified area. This is then classified into either
an "urbanization promotion area", which is an area that has already
been urbanized or is being preferentially urbanized within 10 years, or
an "urbanization control area", which is an area to be controlled. In
Japan, dividing the two is called demarcation, and overseas, it is called
the urban growth boundary. The urbanized area is further designed
to prevent environmental deterioration due to mixed usage by defining
usage areas for residential, commercial, and industrial uses.

Such land use systems have contributed significantly to the healthy
development of cities. However, since the land use plan was established for the existing urban area relatively recently, there are many
existing buildings that do not fit the plan, and in reality, mixed use
is unavoidable. In addition, many cities have realistic land use plans
that do not deviate from the actual situation, so they have had to
follow the status quo rather than pursue ideals. Therefore, there are
many gray zones such as an intermediate area between the urban
area and the suburbs, mixed residential areas, and commercial areas
in urban areas which overlap to form a city.

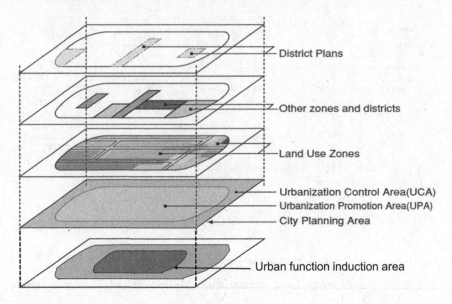

— District Plans

— Other zones and districts

— Land Use Zones

— Urbanization Control Area(UCA)
— Urbanization Promotion Area(UPA)
— City Planning Area

— Urban function induction area

Figure 3.2 Concept of land use planning in Japan.

Source: Modified from Introduction of Urban Land Use Planning System in Japan, MLIT.

Japan reached its peak population in 2008 after a period of high economic growth centered in the 1960s but has experienced a rapidly declining population since then. It is estimated that about 1/4 of the total population will decrease from the peak population by 2050, and the number of vacant homes and land is increasing. Not only this, the maintenance and renewal of aging roads and bridges and other social infrastructure are also a major issue. For local governments, maintenance and management costs are high, but tax revenues have decreased due to the population decline, and the balance between income and expenditure is worsening, which makes city management increasingly difficult. Therefore, how to shrink the urban area expanded during the period of population increase to an appropriate size suitable for the population size is becoming an increasingly important question to answer. The compact city, which was proposed as a sustainable city model, is attracting attention as a potential city model in Japan to combat the declining population.

As a concrete shrinking strategy, the Act on Special Measures Concerning Urban Renaissance was amended in 2016, and a location normalization plan system was created to promote compact city in cooperation with the government and residents and private businesses. Each local government has set up an area where urban

● **Area Division**

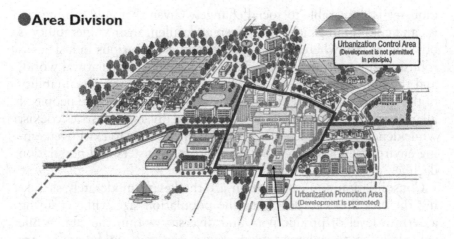

Figure 3.3 Location normalization plan (urban function induction area).

Source: Modified from Introduction of Urban Land Use Palnning System in Japan, MLIT.

functions and residential functions are concentrated in the urbanized area and has begun to induce land use using preferential treatment such as taxation.

3.1.3 Network-type compact city

Since ancient times, many cities have been formed based on growth, both in terms of natural population growth and social growth. The challenge is how to control the increasing urban population while efficiently maintaining urban functions. In a highly mature society, however, the unmarried rate is increasing, the total fertility rate is decreasing, and an aging society with a low birthrate is emerging. In contrast to the conventional urban model that assumes population increase, an urban model that responds to population decline is needed. Unfortunately, there is no urban model that has become popular worldwide in response to the ever-decreasing population.

What are some characteristics of cities that must be maintained even if the population decreases? If this question can be answered, a new city model can be proposed based on the concept. Here, we focus on productivity and diversity as characteristics of cities. This is because ensuring productivity and diversity is the driving force for sustainability. For example, if there is a place to work, people gather and work there, and the population gradually accumulates to form a city. However, the industrial structure that depends on a single industry is vulnerable to social changes. Given that gold mining cities and coal mining cities have risen and fallen, their vulnerability is obvious. A city where various people gather in various industries is robust against social changes. The same is true in the natural world, and the importance of biodiversity is essential for the sustainability of the species. The same is true in cities. In a society where people of all ages and income levels live together, and diverse industries coexist while depending on each other, cities will continue to exist overcoming environmental changes such as natural disasters and population decline.

Consolidation and cooperation are the key to moderately shrinking the city area, according to the population size while ensuring a certain level of productivity and diversity within the city. While consolidating to enhance productivity, the missing functions are supplemented by linking other areas. Focusing on changes in population density, the population density does not decrease uniformly, but rather creates urban areas with certain population densities. Focusing on changes in population density, rather than letting the population density gradually decrease, it is important to maintain a certain value in urban areas while transforming rural areas to idyllic areas for residents using greenery. This type of city model is called a "network-type compact city". A network-type compact city is one in which the various attractive features of the city are aggregated (compacted) in multiple areas that are connected (networked) by various

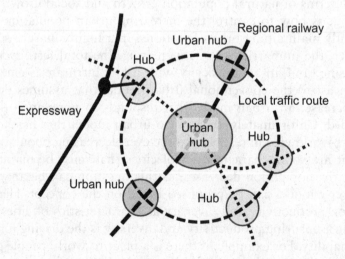

Figure 3.4 A concept of network-type compact city.

modes of transportation, with a focus on highly convenient public transportation.

Although there are many opinions about the genealogy of urban models, it is not a generalized one, but the network-type compact city can be positioned in the lineage of conventional urban models as shown in Figure 3.5. The term "compact city" first appeared in the book "Compact City" (Dantzig and Saaty 1973) that describes a circular city. As a solution to the problems of the past cities that were spread out in two dimensions, a three-dimensional, multilayered city model is proposed to improve the efficiency of urban activities. On the other hand, the compact city as a sustainable urban model originates from the Brundtland Report of 1987, as mentioned earlier. The network-type compact city proposed here is an urban model that inherits the philosophy of the eco-friendly and sustainable urban model and is built mainly through transit-oriented development (TOD), which reorganizes the city through the public transportation network. The diversity of cities pointed out by Jayne Jacobs is an extremely important factor, and it is a concept that underlies the network-type compact city, albeit indirectly.

In addition, there are two types of networks that are used in this city model, a tree type and a rhizome type. Whereas a tree has a hierarchy of trunks, branches, and leaves, a rhizome refers to "a concept in which heterogeneous things that are not hierarchically related to

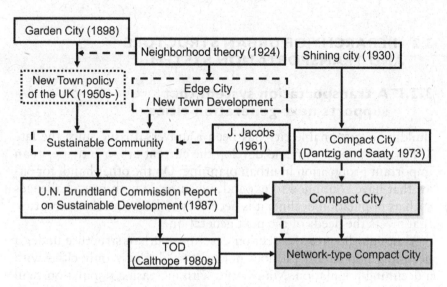

Figure 3.5 A genealogy of network-type compact cities.

each other are connected by a cross-sectional horizontal relationship". Rhizome is a concept developed by the philosopher Gilles Deleuze and the psychiatrist Félix Guattari in "A Thousand Plateau (1980)". Kisho Kurokawa (1987), an architect who used this concept for city comparison, found that Western cities are tree-shaped (hierarchical, trunk, and branch structures), while Japanese cities are rhizome-shaped, which has no center and the characteristics of a complex and constantly changing structure. When a city expands, the tree type is organized and can only expand in a specific direction, while the rhizome type is flexible and can expand in multiple directions.

Focusing on the transportation field, the tree type is desirable if aiming at the overall efficiency of transportation, but from the background of the birth of transportation routes, it can be said that individual demand has developed into a rhizome type. An ideal network can be constructed by combining the tree type, which enables efficient resource allocation, and the rhizome type, which provides the shortest route to the destination. In other words, it is desirable that the network-type compact city is constructed with a multilayered network. The upper layer network becomes a tree type and forms a skeleton while the lower layer network is a rhizome type and each individual is flexibly linked. This is because the upper network emphasizes overall economic rationality, and the lower network emphasizes the consumer's perspective.

3.2 HIERARCHY OF URBAN STRUCTURE AND TRANSPORTATION SYSTEM

3.2.1 A transportation system that supports next-generation cities

Building a sustainable city has been a universal issue since the late 1980s. How to sustain and develop the current urban activities is an important proposition in urban planning. On the other hand, for cities that have continued to expand, when faced with new challenges such as population decline, it is necessary to have an urban structure that meets the needs of the next generation.

Although the previous section describes "urban structure under a declining population", it does not necessarily cover only cities with a declining population. Many cities with increasing populations will eventually begin to decline after they reach maturity. Therefore, it can be said that preparing for the future urban structure from now on is a Leapfrog-type strategy. Here, I will present the shape of the next-generation city and organize the transportation system that supports it.

Next-generation cities can be divided into three levels: metropolitan area level, city level, and district level. At the metropolitan level, the mother city and surrounding cities cooperate with each other to maintain a wide area of living. At the city level, functions such as commerce, housing, and industry in the city are integrated, and they are linked to each other by various transportation means. At the district level, diverse communities are fostered within walkable areas.

When considering a transportation system, how should we combine a tree-type network that aims for overall optimization and a rhizome-type network that is individually optimized to form a next-generation city? First, I would like to consider the transportation system that is optimal for the whole city.

To support the transportation network by car traffic, the required road width depends on the amount of passing traffic. The main road must be wide and requires multiple lanes. The lower the level of the auxiliary main road and the residential roads, the narrower the width and the lower the speed limit during driving. In traffic engineering, such a system is called a traffic cell or traffic zone system.

In urban areas, the hierarchy of such road networks is extremely important for the smoothness and safety of road traffic. Especially in large cities, the traffic volume will increase dramatically, and it will

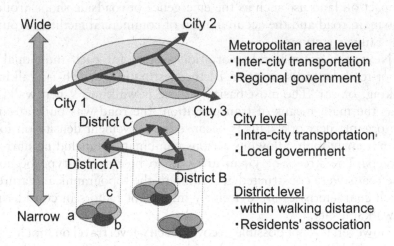

Figure 3.6 Hierarchy and diversity of next-generation cities.

be difficult to develop suitable highways. If the number of vehicles traveling in one lane exceeds approximately 700 vehicles per hour, it will not be possible to maintain smooth traffic. In that case, it will be necessary to develop mass transportation means such as railways and buses. In other words, the transportation network will change from a simple combination of general roads, to a hierarchical road system, to a transportation system that includes public transportation, depending on the transportation demand between the connected locations. Focusing on the capacity of the urban transportation network, the order is as follows:

Railway (subway) > Main road + track system > Main road + bus > Main road > Auxiliary main line > Local road

When introducing public transportation, it is important to devise ways to improve operational efficiency, such as efficient boarding/alighting areas and dedicated driving spaces. In addition, while automobile transportation can provide a flexible door-to-door service, public transportation, such as railways and buses, is a fixed service that connects limited points. Therefore, it is a critical to build a system that allows seamless transfer.

The trunk line system needs to be constructed with the balance of the wide area in mind. Once constructed, it is difficult to change, and the impact is widespread because the traffic flow itself changes. Therefore, it should be decided by a top-down type from a long-term plan or a master plan. In addition, the trunk system has a large impact on land use such as the emergence of roadside shops around the main road and the accumulation of commercial facilities around the railway station.

Next, the optimal transportation system for each individual is door-to-door transportation to the destination, which is walking, biking, or car. The most basic of these is walking. When walking was the main means of transportation, the road was built around an organic network. This is because the movement density on foot is extremely high, allowing various combinations including narrow streets. The streets of Japan are said to be rhizome-type because the roads were constructed according to the topographical features. Such characteristics are extremely important factors in considering a walkable city.

However, it is not possible to cover all city-level travel on foot alone, so it will be necessary to work with other modes of transportation. In other words, in terms of the transportation system, improvement plans should be considered based on the network of branch lines, and it is advisable to update this branch system at the living level of pedestrians. Such a rhizome-type path can be networked with a series of branch line systems up to a certain scale. Reflecting local needs and lifestyles, maintenance and management should be done from the bottom up by the community.

It should be noted that cars support the Rhizome network only in limited places, such as when the number of cars is quite small or there is low density land. Consider this: walking around narrow, winding roads is fun, but is it comfortable to drive around? If cars are flooded in such a city, it is inevitable that traffic will be congested, and accidents will occur frequently throughout the city. This is a common sight in cities in developing countries, where automobiles have suddenly spread in cities built mainly for pedestrians.

In low-density urban areas such as suburbs and rural areas, the traffic volume itself is low, so if safety is taken into consideration, the necessity of hierarchy of roads will be relatively low. In such a place, the network may be a mixture of trunk lines and branch lines.

3.2.2 Transportation facility development and urban development strategy

Looking back on the impact of transportation facility development on the urban structure, it can be seen that when railways and roads are improved, the urban structure changes accordingly. For example, the change in the main means of transportation from railways to automobiles, which began in the early 20th century, can be seen in the differences in the land use distribution model proposed at that

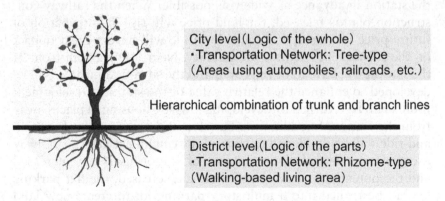

City level (Logic of the whole)
· Transportation Network: Tree-type
(Areas using automobiles, railroads, etc.)

Hierarchical combination of trunk and branch lines

District level (Logic of the parts)
· Transportation Network: Rhizome-type
(Walking-based living area)

Figure 3.7 Conceptual diagram of a two-layer urban structure.

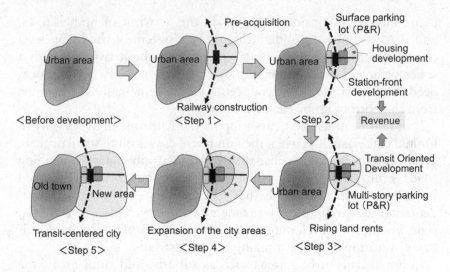

Figure 3.8 Railway construction and urban development.

time. Here, I will present the urban and transportation strategies that lead to a hierarchical urban structure based on experience so far.

Railway development that began in large cities in Japan in the 1910s, or TOD advocated in the United States in the 1990s, has brought about changes in urban land use. The transition of railway development and urban development is shown below.

First, a railway will be constructed around the suburbs of a city. This is because it is not easy to acquire land in an already dense urban area and the high land prices in the city center makes business profitability worse. What is important here is to acquire not only the land necessary for railway construction but also the land around the station in advance as widely as possible. When the railway construction plan is released, the land price will rise in anticipation of future price increases, so acquisition time will have a great impact on the subsequent business profitability. Next, the plaza in front of the station will have to be improved and the surrounding area will be developed to enhance the relative value of the station. Development profits will be returned by developing low-rise housing in places away from the station. In addition, a bus stop and a parking lot for park-and-ride will be developed to expand the catchment area for railway users. After that, when the number of station users has increased and the potential around the station has increased, the flat parking lot can be turned into a multistory parking lot to create new land near the station, where medium and high density development can be carried out. The development of low-rise housing may be carried out relatively quickly to return profits, but it is desirable to carry out urban development after the added value of the land near the station has increased sufficiently. This is because the land can be used at a high level, and the rent increases closer to the station. If such a TOD strategy is successful, its impact will spread to the surrounding areas and the city will expand without strict land use regulations. After a certain amount of time, the city will be completed with the old city and the new city centered around the station in the shape of wings. In many cities in Japan, the distance between the old city center and the station is 2–3 km, which corresponds to the distance from the city center to the suburbs at that time because the old city area was formed within walking distance.

The relationship between railways and urban development extends from stations (one dimension), while roads and urban development extend from linear roads (two dimension). In addition, while most railway stations are new, roads have existed since the era of walking traffic, and land use along the roads is progressing. Therefore, most road improvements are new developments in undeveloped areas or function-updated types such as width expansion based on the alignment of conventional roads. In this way, new road construction and widening are being carried out simultaneously throughout the city, and the impact is widespread and the relationship with urban development is complicated. To simplify the relationship between road improvement and urban development, I will consider the desirable processes from the perspective of new expressway construction and urban development.

First, regarding the relationship between highways and urban areas, the urban planner August Heckscher, in his book "Open Space" (1977), states that in designing highways, the route should never be designed to cross urban areas. He also focused on the barrier effect of highways in the city center, stating that intracity highways should be designed to serve as a circuitous route inside the city center and as a tight bond around the downtown area.

For the development of intercity expressways, routes should be selected avoiding urban areas for the sake of urban conservation and ease of land acquisition. To realize mass transportation and high-speed driving on expressways, signal control is not provided and inflow from the roadside is restricted. Therefore, expressways affect land use around interchanges (IC), which are connection

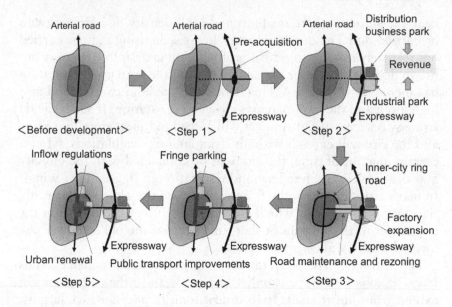

Arterial road

Arterial road

Pre-acquisition

Arterial road

Distribution business park

Revenue

Industrial park

Expressway

Expressway

Expressway

<Before development>   <Step 1>   <Step 2>

Inflow regulations   Fringe parking   Inner-city ring road

Factory expansion

Expressway   Expressway   Expressway

Urban renewal   Public transport improvements   Road maintenance and rezoning

<Step 5>   <Step 4>   <Step 3>

Figure 3.9 Road improvement and urban development.

points with general roads. Desirably, when planning the development of an expressway, it is possible to create a profit return mechanism by acquiring the land around the interchange in advance. The area around the IC on the highway, which is the main artery for truck transportation, is the best location for a distribution business complex. Alternatively, by creating an industrial park, new industries can be brought into the city. Revenue from land sales and rental income are expected for the business entity, and corporate tax and other income are expected if the government is the business entity. If it is not possible to obtain it in advance, it is desirable to formulate a land use plan and regulate land use so that it can be used after opening. While the IC plan is determined, it is necessary to proceed with the development of the main road connecting the neighboring city and the IC. Before the construction of the IC, if the road usually only has a small amount of traffic, the width of the road will be insufficient for the rapid increase in traffic volume, which will cause chronic congestion. If that is the case, a road widening project that matches the new road alignment and a land readjustment project that matches the roadside land use can be implemented. To eliminate through traffic to the city center, the city center ring road can also be improved. It is particularly difficult to widen the road as it gets closer to the city area, and therefore it is necessary to take measures linked to city planning projects such as land replotting and multilevel replotting. Once the road network begins to expand, the public transport network in urban areas should also be improved. Priority roads for public transportation and dedicated driving lanes should be provided on major arterial roads to improve the convenience of public transportation. In addition, fringe parking should be provided at the intersection of the ring road and the radial road to promote the switch from cars to public transportation. By constructing a certain road network, through traffic to urban areas can be eliminated, and the surplus space can be reorganized for public transport and pedestrian traffic. When both the road network and the public transportation network are completed, the inflow to the city center will be restricted, and the city can be guided toward a more pedestrian-friendly city encouraging city center regeneration.

In general, if city planning is based on railway development, the strategy is to begin development from the inside of the city so that the station will eventually become the center. On the other hand, when planning a city based on road improvement, the strategy is to begin from the outside and work inward toward the old city. The process shown here is just an example of various urban and transportation strategies. The strategy to be used should be decided based on the topography of the city, facility layout, and the progress of motorization. The important thing is to create a strategy on the time axis, assuming that land use will always be affected by traffic improvement.

The renewal of the city center, which was mainly developed in the age of walking, should also be carefully considered. The city center is also a social urban heritage with hundreds of years of history and culture, and strongly inherits the spiritual image of the city. Streets built on the premise of pedestrians are narrow and often nonlinear. To reorganize such places to facilitate automobile traffic, roadside facilities will have to be dismantled or relocated. Sufficient debate is needed as to whether it really makes sense to renew the core of the city through redevelopment, or whether it makes more sense to preserve historical values. The city center, if destroyed to suit the needs of one generation, cannot be restored. From this point of view, the proposal presented here has the intention of preserving the city center as much as possible and regenerating it as a pedestrian-friendly city. It is natural that a city created during the pedestrian era is a city that prioritizes pedestrians.

REFERENCES

George B. Dantzig, Thomas L. Saaty. 1973. *Compact City: A Plan for a Liveable Urban Environment*, San Francisco, CA: W. H. Freeman & Co.

August Heckscher. 1977. *Open Spaces: The Life of American Cities*, New York, NY: Harper & Row.

Kunihiro Kamiya. 1983. *Sociology of Urban Comparison; A Comparison of Prototypes in European and Japanese Cities*, Kyoto: Sekaishiso-sha (in Japanese).

Kisho Kurokawa. 1987. *The Idea of Symbiosis – A Lifestyle of Living in the Future*, Tokyo: Tokuma Shoten (in Japanese).

Akinori Morimoto, 2019. Future City Planning with Autonomous Vehicles, *The "Jutaku" A Monthly of the Housing*, 68:14–18 (in Japanese).

Yasuo Nishiyama. 2002. *What Is Japanese Urban Planning?* Kyoto: Gakugei Shuppan-Sha (in Japanese).

Barrie Shelton. 2012. *Learning from the Japanese City: Looking East in Urban Design*, 2nd ed., London: Routledge.

Chapter 4

Land use and transportation

4.1 INTERRELATIONSHIP BETWEEN LAND USE AND TRANSPORTATION

4.1.1 Land use and transportation

The relationship between land use and transportation is often compared to the relationship between chickens and eggs. That is because land use changes traffic, and traffic changes land use. In the era when walking was the main means of transportation, houses and facilities were concentrated within walking distance. Later, when railroads were laid to allow long-distance travel, people's living spheres expanded significantly along the railroad network. The center of the city was gradually drawn to the railway station and a railway-type city was formed. In the 20th century, when the use of automobiles became widespread among the working people, cities expanded in all directions along the roads. Land use was released from the constraints of walking distance, and the direction constraint of the railway network, and will expand two-dimensionally as long as there is a road.

It is easy to understand that road construction will expand the city, but this is often overlooked in actual policy. Such signs can be seen, for example, the construction of a bypass road or a ring road that detours around the city center, which was devised as a countermeasure against traffic congestion in the city center. Traditionally, roads have been formed to connect cities, and as automobile use increases, the city center suffers from significant road congestion. This is because even through traffic, which is originally useless in the city center, flows into the city center. One of the preventative measures is the development of a bypass road to eliminate passing traffic. The effect of the bypass road is enormous, and the passing traffic that flows into the city center is greatly reduced, achieving the purpose of alleviating traffic congestion. However, roadside shops are lined up along the bypass roads in the suburbs, and large-scale parking lots can be easily constructed, so large-scale commercial facilities will gradually accumulate. In large cities, the demand in the city center is high, so it is possible to achieve a certain balance by dividing the roles with the suburbs. However, in small and medium-sized cities, the central part of the city loses, and the suburbs become bloated. The original purpose of alleviating traffic congestion in the city center was to eliminate transit traffic, but in turn promoted the suburbanization of facilities that were previously located in the city center, and reduce the traffic demand toward the city center. Thus, although the purpose of the transportation plan was achieved, the city plan can develop a new problem of decline in the city center.

If a city becomes overly dependent on automobiles, urban problems such as the decline of the city center, and an increase in traffic accidents, environmental problems, and the number of vulnerable people can start to affect its residents. If the problem becomes so serious that it cannot be overlooked, certain restrictions will be required to manage the use of automobiles. There are many theories that have been used to examine the links between transport and urban development, but D. Banister (1995) states that the most important one has been classical location theory, which is based on assumptions from land economics of optimality and equilibrium in land allocation. In

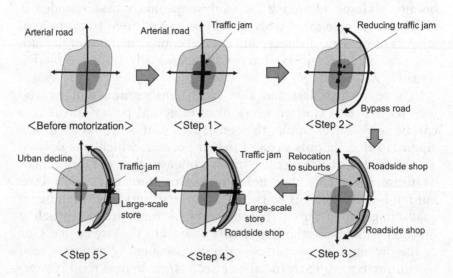

Figure 4.1 Impact of bypass road development.

this chapter, I will consider ways to improve the balance between land use and transportation, while keeping the theoretical framework in mind.

4.1.2 Factors forming automobile-dependent cities and their countermeasures

To return excessive automobile use to an appropriate level, it is necessary to understand the cause of why automobile-dependent cities have expanded so far. The causes can be summarized in the following five reasons.

1. Factors related to land use: convenience of transportation is an indispensable requirement for living in the city. Therefore, the place of residence and workplace is selected in consideration of the traffic environment. When the car usage environment is improved, suburban land with abundant greenery and low land prices is preferred to the city center, which is crowded and has high land prices. As the number of car passenger increases, the density of cities gradually decreases, causing a relative decline in the city center.
2. Traffic-related factors: as users shift from walking or public transport to automobiles, the number of public transport users will decrease. To maintain public transportation, transportation companies will abolish unprofitable routes and reduce the frequency of operations on routes with fewer users. A decline in public transport service levels will lead to a further decline in public transport users.
3. Life-related factors: in the process of motorization in many cities, the wealthy first get a comfortable way to travel by owning a car. This establishes the car as a status symbol. As the economy grows and people's income levels rise, car ownership explodes. Once people own a car, traffic behavior is personalized to meet various needs and the lifestyle itself changes. Diversified demand cannot be met by means of transportation other than cars, and cars will become a part of our life.
4. Economic factors: cars are expensive, but the additional cost of using it after purchase is relatively low. Additionally, there is no additional labor cost to pay if the buyer drives. An appropriate cost-sharing system is inadequate for external diseconomy such as traffic congestion and environmental

deterioration. Taxes such as gasoline tax are levied to maintain the road infrastructure, but the price is relatively low. Some countries are taking tax incentives as a national policy because the activation of transportation activities is linked to the economy.

5. Political factors: congestion becomes a social problem when many people use cars. Congestion significantly reduces urban activity, and its elimination tends to be a political objective. Alternatively, road improvement may be requested to revitalize the urban economy and promote industry, which is also likely to become a political issue. Through political judgment, road construction may be implemented even in areas with low demand, and the development of road networks will induce further vehicle use.

In this way, the dependence on automobiles encourages the further automobile use with the five factors intricately intertwined. Of course, a certain degree of dependence on automobiles is desirable because it stimulates urban activities. However, when the dependence exceeds a certain level, it causes various urban problems. The countermeasures for each of these factors are as follows.

1. Land use: suburban location regulation, city center revitalization
   Land use restrictions are implemented using the city planning system to curb chaotic suburban development. Undesirable development can be curtailed by enforcing regulations on large-scale development and inappropriate uses. Alternatively, as a measure to revitalize the city center, administrative subsidies and deregulation can be implemented for redevelopment projects in the city center.

2. Transportation: fostering public transportation, developing alternative transportation methods
   Administrative subsidies can be provided to maintain declining public transport. Alternatively, on behalf of the withdrawn private business, the government can take the lead in managing public transportation through administrative spending. In addition, the subsidy system for new public transportation projects can be enhanced.

3. Life: changes in values (away from the car), walking due to health consciousness

Values for car ownership are changing, especially among young people. Cars have changed from what they own to what they use and car sharing services have been created to support this. In addition, due to growing health consciousness and environmental consciousness, policies are being taken to give preferential treatment to people and environment-friendly transportation such as bicycles and walking.

4. Economy: taxation of automobiles, lowering of public transportation.

The appropriate taxation on automobiles can also be considered. For example, road pricing, which charges people appropriately for the use of congested roads. Alternatively, setting a low fare for public transportation to increase the number of users.

5. Politics: global environmental policy, sustainable development

In response to common policy issues in the world such as prevention of global warming, public transportation policy can be regarded as a part of environmental policy. Bicycles are increasingly being reviewed as an environmentally friendly transportation option, and more and more countries are focusing on bicycles in their urban transportation policies. Therefore, various policy decisions can be carried out to support sustainable development.

4.2 INTEGRATION OF LAND-USE PLANNING AND TRANSPORTATION PLANNING

4.2.1 Relationship between land-use planning and transportation planning

To realize an ideal city in the future, it is necessary to consider the fusion of land-use planning and transportation planning. Although its importance has been pointed out as a planning theory, its realization involves difficult problems. This is because, as Ralph Gakenheimer (1993) points out, land-use planning, and transportation planning are fundamentally different from numerous perspectives. First, looking at the scales of concern, the land-use plan is limited to a certain area, while the transportation plan targets a wide area network. The objectives of land-use planning are diverse, such as hygiene, comfort, safety, and revitalization, but transportation planning is relatively simple, such as smoothness and safety. The forecast period of the land-use plan is relatively short, but the transportation plan is often a long-term forecast. In addition to this, characteristics such as analysis method, government involvement, prospects for implementation, and implementation unit are significantly different.

While there are many differences between land-use planning and transportation planning, they do influence each other. Work oriented transportation activities are created as workers work away from their homes and transportation facilities such as roads and railroads are developed to support these activities. When roads are improved for the purpose of alleviating traffic congestion, the location potential along the road will increase and urban facilities will be developed. When a new facility is created, new urban activities will occur there. Land-use planning will affect transportation planning, and the results of transportation planning will require new land-use planning. If urban activities and transportation activities are regarded as urban demand, urban facilities and transportation facilities can be understood as supplies that support the demand. The former is in the category of social planning, and the latter is mainly the subject of physical planning.

In a land-use plan, the realization of the plan induces transportation activities and puts a load on transportation facilities. Therefore, it is necessary to thoroughly check and formulate a land-use plan that does not exceed the capacity of transportation facilities. The important thing is that if the demand for transportation exceeds the supply of transportation facilities, a new transportation plan must be made, and the transportation facilities must be improved accordingly. As mentioned above, the improvement of transportation facilities will stimulate new land demand,

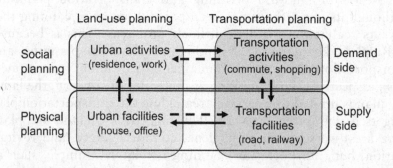

Figure 4.2 Interrelationship between land-use planning and transportation planning.

so there is a high possibility that the situation will be different from the original land-use plan.

From the perspective of transportation planning, when a transportation plan is made with a land-use plan as a given condition and the transportation facility is improved accordingly, the transportation facility itself generates new land demand, which is different from the initial prerequisite. The more roads are built to alleviate traffic congestion, the more convenient the car will be, the more total traffic there will be, and as a result, congestion cannot be alleviated. Such a phenomenon appears in many areas and is called "induced traffic". Even more annoying is that land use itself is likely to change significantly.

4.2.2 Population density and traffic density

The population density of large cities was high when walking was the main form of transportation. Since railroads were introduced after the 19th century, walking was the main means of transportation in cities in the 18th century. Edo (now Tokyo), one of the largest cities in the world, has a population density of 23,000 people per km². That means it is probable that the streets at that time could accommodate a living density of about 23,000 people per km². On the other hand, looking at the population density of cities mainly built around railways, the city with the highest population density in Tokyo, which has the world's largest railway network, is Toshima Ward with about 21,800 people per km². It is interesting that the population densities of both walking and rail cities are similar. The population density of the 23 wards of Tokyo is 15,100 people per km².

On the other hand, looking at cities that mainly use bus rapid transit (BRT), the population density of Curitiba is about 4,000 people per km², and that of Bogota is about 6,000 people per km². In addition, the population density of Houston, the city with the highest gasoline consumption per capita in the United States, in other words, the city with the highest dependence on automobiles, is about 1,400 people per km².

According to NACTO, in order to move 10,000 people in one hour, pedestrians and bicycles need only one lane (about 3.5 m wide), buses need two lanes, and cars need 13 lanes. Incidentally, it is estimated that to transport 10,000 people in a flying car, 38 lanes would be needed as a landing area.

From these data, a rough estimate of the relationship between the maximum population density of a city based on the primary mode

What Does It Take to Move 10,000 People Per Hour?

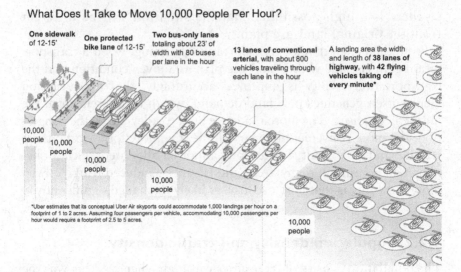

Figure 4.3 Transportation and road space.

Source: NACTO, Blueprint for Autonomous Urbanism, second edition (https://
nacto.org/publication/bau2/).

of transportation, and the maximum traffic density of that mode of transportation itself can be deduced. As a characteristic tendency, the railway and walking have the largest density, while buses have half the value, and a car has a tenth in value. In short, both population density and traffic density show the same tendency, and both indicators have a linear relationship (Table 4.1).

In addition, since an actual city is made up of a balance of various modes of transportation, it is natural that a comprehensive

Table 4.1 Population density and traffic density

Traffic means	Railway	Walk	Bus	Car
Traffic volume (person/ hour/lane)	10,000	10,000	5,000	1,000
Traffic density (person/ sec/m)	0.8	0.8	0.4	0.08
Population density (person/km^2)	20,000	20,000	10,000	2,000

evaluation is necessary. A comparison of population density and traffic density shows that high density transportation, such as walking and public transport, is essential for maintaining a densely populated city.

4.2.3 Which comes first, land use or transportation

Focusing on the relationship between land use and transportation, the question "Which comes first, land use or transportation improvement" will be considered. If urban land use is progressing and traffic is to be changed later, the transportation company has users from the time of the opening and can expect a certain amount of transportation demand. Profitability will improve if fares are collected from many users, but the increase in added value of the surrounding land caused by the development of transportation facilities will be absorbed by the former landowners in the surrounding area.

In order to return the development profit to the transportation business, it is necessary to acquire the land in advance, but the land price is high because the land use is already progressing, and it is difficult to reach a consensus on the land acquisition. Improving transportation to the most prosperous city center will be an extremely difficult project in terms of business profitability and consensus building. An example of this can be seen in the construction of the Yamanote Line, which runs through the heart of Japan's capital, Tokyo. The Yamanote Line (34.5 km long), which runs in a loop through the center of the city, serves as a hub of the public transportation network connected to the

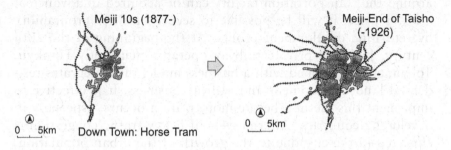

Figure 4.4 Construction of a ring railroad in central Tokyo, the Tokyo Yamanote Line.

Source: Ministry of Land, Infrastructure and Transport (Urban Transport Facilities in Japan, 2002, p. 1–9).

suburban railroads and the subway that runs through the center of the city. In history, the last opening on the Tokyo Yamanote Line was between Ueno and Tokyo, which was the largest downtown area at the time. This difficult project was carried out as a reconstruction project for the disappearance of the central city area due to the Great Kanto Earthquake (1923), and the current circular operation was realized. In other words, it would have taken even longer to be realized without the earthquake.

The initial investment for a transportation facility development is large. However, the cost will be covered by financial resources based on toll road usage fees, fares, and gasoline taxes paid by direct users, as well as indirect business taxes and other taxes. Since there is a wide range of beneficiaries from public utilities such as road and railroad construction, construction costs are disbursed from a variety of financial resources. But beneficiaries along newly constructed roads and around stations do not always bear the costs that commensurate with the benefits. In economics, accepting benefits without paying a reasonable price (cost for supply) is called a "free ride". This problem is difficult to solve, especially for public goods such as roads, which are non-excludable (non-paying consumers cannot be prevented from accessing it).

On the other hand, in the case of transportation, the demand for transportation can be low for a while even if the transportation facilities are put into service, and in the case of railways, it is difficult to reimburse the construction cost, and the fare income required for operation cannot be expected. In the case of road construction, there is a high possibility that public opinion will criticize it as a wasteful investment. However, if the land around the transportation facility can be acquired in advance at a lower price, it will be possible to secure business profitability by returning development profits. At the beginning of the 20th century, Japanese private railway operators (currently Hankyu, Tokyu, etc.) developed with a business model that integrates residential land development and railway business. It is effective to implement this method before motorization occurs, especially in developing countries. Alternatively, if heavy traffic congestion is expected in the city due to the growth of the urban population, the attractiveness of railways that can run on time will increase, and the demand for residents along the railway will increase. By combining attractive transit-oriented development (TOD) and railway business, the urban structure can be guided appropriately.

4.3 POLICY TO ENSURE CONSISTENCY BETWEEN LAND USE AND TRANSPORTATION

4.3.1 Land use and transportation integration policies in city planning

Land use and transportation are the most basic elements in city planning, and methods for achieving consistency is basically prepared in the city planning system. This system is a two-stage city plan, which is a high-level plan that balances the whole and a low-level plan that involves specific regulatory guidance.

In Germany, the Federal Construction Law enacted in 1960 introduced a two-stage city planning system, the F plan (Flächennutzungsplan) which shows the basic policy of land use for the entire city area, and the B plan (Bebauungsplan) which is legally binding according to the F plan. The basis of French city planning is a two-tiered planning system, the Regional Integration Plan (SCOT) and the Local City Planning Plan (PLU). SCOT is a wide-area plan formulated by a wide-area administrative organization while PLU is a detailed plan that limits individual development activities, and directly limits private rights. In the United Kingdom, it consists of a structure plan, which is a high-level plan created by the county to show future development strategies, and a local plan, which is a guideline for considering development permits created by the district. The city planning system in the United States varies from state to state, but basically the municipalities have the authority to plan the city. The city basic plan is called a general plan, a comprehensive plan, or a master plan, and both are voluntary plans. Development control is mainly carried out by zoning system and site subdivision control. In Japan, the Urbanization Promotion Area and Urbanization Control Area were newly established by the New City Planning Act of 1968, and the basics of the land-use regulation system were established. Subsequently, in 1992, the Municipal Master Plan showing the future vision of the city was established, and the framework of the two-stage city planning system emerged.

Although the system and effective power differ depending on the country, in general, wide-area coordination with other cities and coordination with other fields such as the transportation are made in the upper plan, and legal permission is given to the location entity in the lower plan based on that. With this system, land use and transportation are balanced by allowing high-density development along the main roads and increasing the development capacity around railway stations.

4.3.2 Direct and indirect public intervention

Urban planning techniques for adjusting land use and transport inconsistencies include direct and indirect public intervention. Direct public intervention refers to the integrated development of a certain area by the developer based on city planning. Indirect public intervention is a land use regulation that defines the type and shape of land use in the city plan in advance and is effective when the location entity carries out the rebuilding.

A land readjustment project is a typical method of direct public intervention. The Lex Adickes, enacted in Germany in 1902, created a system to provide a certain percentage of the construction area as public land free of charge. With reference to this system, Japan established a land readjustment project to efficiently improve public land such as roads by arranging land lots and increasing added value. This method of withdrawing project costs by raising land prices has been used in the reconstruction of urban areas burned down in the Great Kanto Earthquake (1923) and in the subsequent war damage reconstruction projects, and has been implemented in many cities since then.

The mechanism is as follows. Landowner A owns irregular land with narrow streets, so the property value of the land is low. By integrally developing the surrounding land, the shape of the land is adjusted, and sufficient roads and parks will be developed. As a

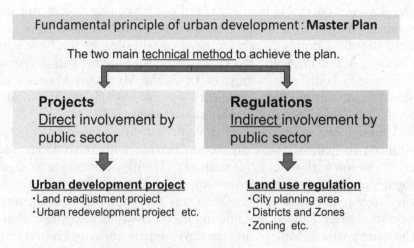

Figure 4.5 Technical process of urban planning.

result, the asset value of the land of landowner A will rise, and the land commensurate with the increase will be contributed as public land. Although the area owned will be reduced, the asset value of the land will remain the same as before (the principle of correspondence). The executors of land readjustment projects can create public land without paying a large amount of money to acquire the land of the landowner. As a rough estimate, if the land price increases by 10% due to the project, about 9.1% (= 1 − 1/1.1) of land can be acquired free of charge.

When the urban area is dense and it is difficult to replace the land on a flat surface, new surplus land is created in the upper space by three-dimensional replotting, and the project cost is generated by selling it. This is called an urban redevelopment project. At this time, if there is a system that can increase the legal upper limit of the development of floor area (floor-area ratio [FAR]) by setting back from the boundary line of the site and providing public space, road space can be created by this incentive.

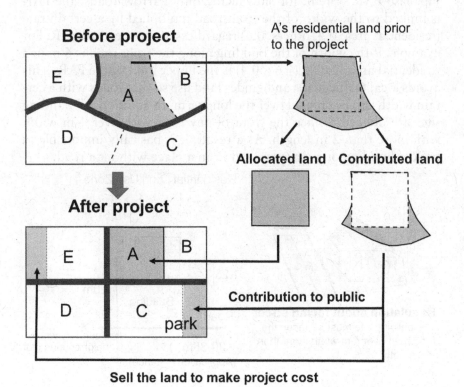

Figure 4.6 Mechanism of land readjustment project (conversion of rights).

Indirect public intervention is a widespread regulation and one of the long-term planning techniques. It consists of dividing the land into areas for urban land use and areas for maintaining the natural land use such as farmland and forests. The distinction between the two is called the urban growth boundary. By setting areas that guide or curb development according to the growth rate of the city, it is possible to maintain a balance between land use and transportation at the city level. Furthermore, land use in the city is subdivided by zoning, and each site has an upper limit on the total floor area of the building (FAR) and an upper limit on the building area (Building Coverage Ratio, BCR). This maintains the relationship between land use and transportation at the site level. Examples include FAR regulation (form regulation), a regulation about facing the street, and slant plane restrictions.

The FAR indicates the legal upper limit of the total floor area that can be built on the site, but due to the relationship with the front road, the upper limit cannot be reached. For example, under the Japanese FAR system, for sites facing only narrow roads, the FAR is limited to the width of the front road multiplied by a certain rate (residential area: 0.4, others: 0.6) regardless of the specified FAR. For example, if the width of the building along the front road is 6 m in a residential area, 240% (= 6 × 0.4) is the upper limit of the FAR. This avoids local traffic loads and guides land use so that roads with a certain width can be created over the long term. In addition, all building sites are obliged to face the front of any road with over 4 m width with more than 2 m length. As a result, it is basically impossible to build a building on a narrow road or in a place without a road.

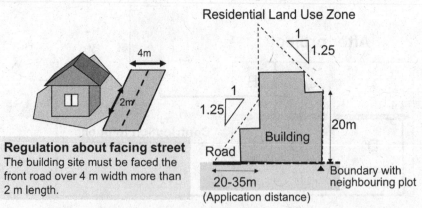

Residential Land Use Zone

1
1.25

1
1.25

20m

Building

Road

Boundary with neighbouring plot

20-35m
(Application distance)

4m

2m

Regulation about facing street
The building site must be faced the front road over 4 m width more than 2 m length.

Figure 4.7 Restriction on floor-area ratio according to the width of the adjoining road.

Although direct public intervention can solve the problem effectively, there are issues such as consistency with areas other than the target area and securing of project costs. Indirect public intervention is a regulatory guide to market trends and can be widespread with little additional cost, but it will take many years to achieve. If the system can implement strong land use regulations, it is possible to control traffic demand to some extent. For example, J. Pucher and C. Lefevre (1996) points out that the stronger the land use regulations in each country, the lower the dependence on automobiles.

The ideal planning method is to first create a high-level plan based on a long-term urban perspective, and then set land use regulations such as growth boundaries and zoning. In order to support and realize this, direct public interventions such as urban planning projects are to be implemented for areas with poor environment and high development potential.

4.3.3 Traffic assessment

Traffic impact assessment (TIA), or simply traffic assessment, is a method and system for assessing in advance the effects of development plans on transportation in order to implement traffic policy from the perspective of the harmonizing land use and transportation. According to the Institute of Transportation Engineers, traffic impact studies project future transportation demands, assesses the impact of changes in demand, and suggest ways to mitigate the adverse effects of land use changes in defined geographic areas.

Its origins can be traced back to the United States in the 1980s, when federal subsidies were reduced, making it difficult for governments to improve roads. In the case of causing road congestion, the developer paid a reasonable burden to internalize the external diseconomy. In US traffic assessments, the results of the traffic assessment are compared with the standards for transportation infrastructure development, and if this is not satisfied, the developer side is burdened with the transportation infrastructure development equivalent to the traffic impact on the surrounding area, or the development plan is revised so as not to cause an impact. When requesting the burden of transportation infrastructure development from the developer side, there are cases where the transportation infrastructure development itself is requested and the traffic impact fee is required.

The implementation of traffic assessment was institutionalized in the 1980s in Western countries. The application in the United States varies from region to region with different standards, but a traffic impact study was recommended to be conducted whenever a proposed development will generate 100 or more added peak direction trips to or from the site during the adjacent roadway's peak hours or the development's peak hours. In Japan, the application of the system began in the 1990s, and since 2000, large-scale retail stores with a commercial floor of 1,000 m^2 or more are obliged to carry out traffic assessments. In conducting a traffic assessment, it is extremely important to grasp the traffic situation quantitatively and in detail. Due to innovations in information technology since the 1990s, computer processing power has been dramatically improved allowing signal control, lane composition, etc. to be reproduced, and visual expressions to be expressed by animation. This has allowed the analysis of traffic conditions to be conducted faster and more efficiently.

By visualizing the impact of land use on traffic and directly linking development (cause) and traffic congestion (results), development plans that do not cause traffic congestion and road maintenance that matches development are implemented on a project-by-project basis. By conducting a traffic assessment, it is possible to secure an appropriate number of parking lots, determine the location of parking lot entrances and exits, install additional lanes, improve intersections, change store visit routes, and introduce public transportation.

4.4 TOWARD NEW LOCATION GUIDANCE MEASURES

4.4.1 Location management

In the period of population growth, land use progresses ahead of other areas, so the method of regulating and guiding development trends is effective. By controlling the increasing land demand in urban areas, it is possible to get closer to the city depicted in the master plan. On the other hand, since the land market is stagnant during the period of population decline, the rate of increase in land prices due to the land readjustment project is low, and it is difficult to obtain development profits from urban development. In addition, regulations that set development limits, such as the FAR system, are less effective when the population is declining.

In such cases, it is difficult to guide land use to the desired form only through direct public intervention such as city planning projects and indirect public intervention such as land use regulation. Therefore, we will return to the basics of land use and transportation mechanics and propose a new location guidance measures by considering the relationship between the two. Here, we will change the perspective of city planning from a land-use-first strategy to a transportation-first strategy. If shrinking a low-density urban area is a sustainable urban structure, a strategy to implement transportation facilities to guide it to the ideal urban structure can be proposed. In other words, instead of improving transportation facilities to facilitate traffic, we will improve transportation facilities to optimize land use.

The location guidance planning technique that supports the integration of land-use planning and transportation planning is called Location Management (LM). LM is defined as "a technical system that guides the locations of housing, commerce, industry, etc. that make up a city to the right place for the purpose of forming a sustainable city". Specifically, LM consists of the following three policies. Here, residence guidance will be described as an example.

1. Pull measures: this is a residential guidance measure on the aggregation side, which indicates measures to induce the migration of residents into the area to be systematically aggregated. Developing transportation facilities to increase the land potential in the centralized area. In particular, the introduction of next-generation public transportation is effective. In response to such transportation strategies, complex policies such as TOD, urban rent subsidies, housing attachment obligations, and infill development can be implemented.
2. Push measures: this is a move-out measure that shows the technology for returning the residential land after relocation to nature and the measures for reuse. This corresponds to mitigation and adaptation in the environmental field. For example, conversion of land after withdrawal to a natural state such as green space (Convert), rental of vacant houses to younger generations (Reuse), reverse delineation (Regulate) and reduction (Reduced) can be listed.
3. Relocation measures: measures to induce residents on the withdrawal side to relocate to the consolidation side. While the above two measures do not limit the origin or destination, this measure is characterized by identifying the combination of move-in and move-out. Examples of this measure include additional relocation subsidies for moving from suburban areas and recapture of asset value by moving in from areas with inconvenient public transportation (Location Efficiency Mortgage System).

Table 4.2 Land use and transportation strategies

	Land use	Transportation
Positive strategies	Mixed land use Comprehensive development Transit oriented development Incentive zoning etc.	Transit facilities improvement Increase of LOS Related facilities improvement etc.
Negative strategies	Restrict the development outside the planning Planned site design Restrict zoning control etc.	Higher parking charge Discourage the car use Appropriate pricing to car use etc.

Note: Four strategies are required for improvement.

4.4.2 Land use and transportation integration strategy

There are two perspectives in developing an integrated land use and transportation strategy. Positive measures which support specific markets to achieve their goals, and negative measures, which restrains opposing markets. It is effective to implement both at the same time, and it is known as a carrot-and-stick policy (Table 4.2).

The measures to increase the use of public transportation in a society with excessive dependence on automobiles are as follows. First, as for measures related to land use, there are measures to increase the development density by relaxing height restrictions and FARs by using mixed land use and incentive zoning. Alternatively, developments that assume the use of public transportation, such as TOD, are all measures that support the increase in public transport users. In addition, land use measures to curb the use of vehicles include restraining development outside the planned area, site design that eliminates passing traffic, and strict land use regulations. As for transportation-related measures, the introduction of attractive public transportation and the improvement of public transportation service levels will increase the use of public transportation. A system that raises parking fees in the city center or charges cars that flow into the city center will also reduce the use of cars. For measures located in four quadrants consisting of a combination of land use and transportation, and positive and negative effects, at least one or more measures should be taken from each quadrant and

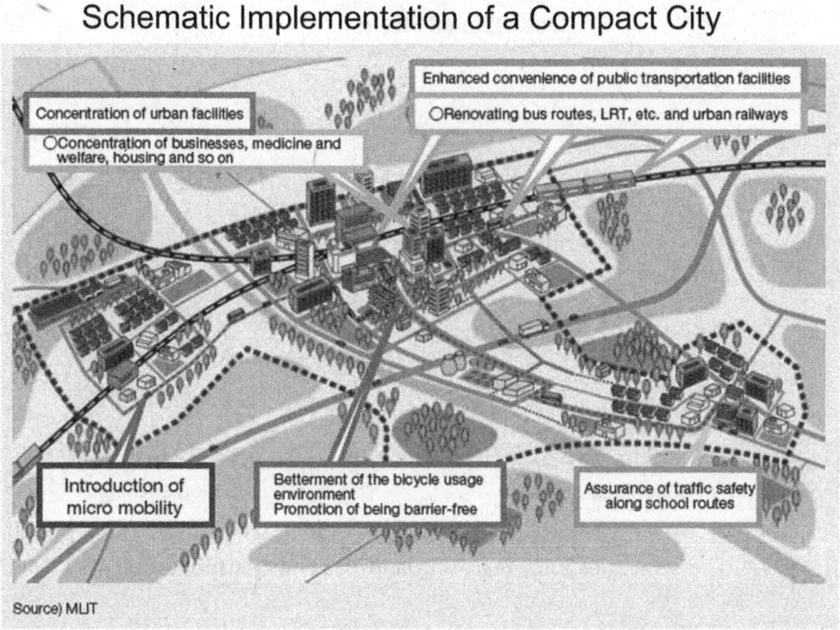

Schematic Implementation of a Compact City

Figure 4.8 Forming a compact city.

Source: White Paper on Land, Infrastructure, Transport, and Tourism in Japan (2012).

an effective combination should be considered. That is the integrated strategy of land use and transportation.

The following measures are an example of an integrated strategy of land use and transportation for an intensive urban structure. While tightening location restrictions in the suburbs, urban functions such as offices, hospitals, welfare facilities, and housing will be concentrated in the city center. In addition, the convenience of public transportation facilities will be improved, and bus services will be improved, or next-generation public transportation will be introduced according to the characteristics of the city. In conjunction with this, the introduction of micro-mobility for intra-district will support short-distance travel.

By performing detailed management of land use where traffic occurs according to regional characteristics and the needs of the times, it is possible to achieve consistency between the transportation strategy and the land use strategy. Since there are uncertainties in the manifestation of the impact from traffic to land use and from land use to traffic, it is important to implement a management cycle

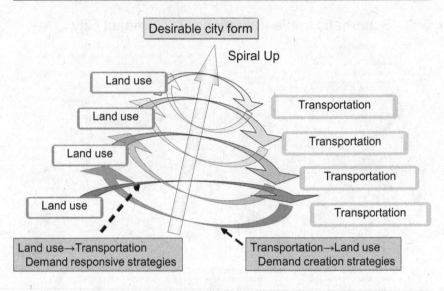

Figure 4.9 The strategies for integrating land-use and transportation.

on a time axis for a certain period. The PDCA cycle is a typical management cycle and is an acronym for Plan (P), Do (D), Check (C), and Action (A). It has been developed as a continuous improvement method such as quality control in production technology.

In this section, the relationship between land use and transportation was expressed by an integrated strategy rather than a comprehensive plan. This is because plans generally have a strong static aspect of drawing ideals for the target year, while strategies assume that they will respond dynamically to changes in the environment. The land market is particularly active during the period of population growth, and the rate of progress in land use is rapid, so transportation facilities will follow suit. Transportation facility development is a "demand-responsive strategy", such as when resolving traffic congestion in the short term. On the other hand, since transportation facility development will change future land use, transportation facility development to guide to ideal land use can be said to be a "demand creation strategy". In this case, the onset of the effect may be medium- to long-term. By skillfully using both of these strategies, the city can be guided to the ideal city.

REFERENCES

David Banister. 1995. *Transport and Urban Development*, London: E&FN Spon.

City and Regional Development Bureau and Building Research Institute, Ministry of Land, Infrastructure and Transport. 2002. *Urban Transport Facilities in Japan 2002.*

Roger L. Creighton. 1970. *Urban Transportation Planning*, Urbana, IL: University of Illinois Press.

Ralph Gakenheimer. 1993. Land Use/Transportation Planning: New Possibilities for Developing and Developed Countries, *Transportation Quarterly*, 47(2):311–322.

Institute of Transportation Engineers (ITE). 1992. *Traffic Engineering Handbook*, New Jersey: Prentice-Hall.

Institute of Transportation Engineers (ITE). 1994. *Manual of Transportation Engineering Studies*, New Jersey: Prentice-Hall.

Institute of Transportation Engineers (ITE). 2006. *Transportation Impact Analyses for Site Development*, New York: ITE.

Mitsuhiko Kawakami. 2012. *City Planning*, 2nd ed., Tokyo: Morikita Publishing Co., Ltd. (in Japanese).

Akinori Morimoto. 2014. Urban Planning and Transportation in the Future: Planning and Implementation of Urban and Transportation Strategies, *The Journal of the Land Institute*, 22(1):1–6 (in Japanese).

Akinori Morimoto. 2015. Transportation and Land Use, In *Traffic and Safety Sciences: Interdisciplinary Wisdom of IATSS*, 22–30, Tokyo: International Association of Traffic and Safety Sciences.

Akinori Morimoto. 2019. Advanced Transport and Compact City in Local Cities, *Urban Renewal*, 594:41–44 (in Japanese).

National Association of City Transportation Officials (NACTO). 2019. *Blueprint for Autonomous Urbanism*, 2nd ed. https://nacto.org/publication/bau2/ (accessed August 15, 2020).

John Pucher, Christian Lefevre. 1996. *The Urban Transport Crisis in Europe and North America*, London: Macmillan Press Ltd.

Tatsuya Seki, Akinori Morimoto. 2010. A Review of Traffic Impact Assessment in Japan, *Journal of Japan Society of Civil Engineers D*, 66(2):255–268 (in Japanese).

Kazumasa Suzuki, Akinori Morimoto. 2011. A Research to Systematize the Measures of Location Management Toward the Compact City, *Journal of Japan Society of Civil Engineers D3*, 67(5):I_315–I_320 (in Japanese).

Randall Thomas. 2003. *Sustainable Urban Design: An Environmental Approach*, London: Spon Press.

Shun-ichi Watanabe. 1993. *The Birth of "Urban Planning" – Japan's Modern Urban Planning in International Comparison*, Tokyo: Kashiwashobo Publishing Co., Ltd. (in Japanese).

Takashi Yajima, Hitoshi Ieda. 2014. *Tokyo, a World City Created by Railroads*, Tokyo: The Institute of Behavioral Sciences (in Japanese).

Chapter 5

Consider transportation based on the city

5.1 WHAT IS DESIRABLE TRANSPORTATION?

5.1.1 Transportation as derived demand

When a person moves to a destination, the activity performed at the destination is the "primary demand", and the movement itself is the "derived demand" that arises from the original purpose. Commuting to work, going to school, and traffic on the way to the shops are all derivative demands. Since transportation itself is not the purpose, it is desirable that transportation be carried out quickly and safely at a lower cost. If travel time is reduced, it will result in reduced costs, reduced risk of traffic accidents, and reduced burden of travel (negative utility). The ultimate transportation technology would be teleportation because the negative utility disappears when the travel time becomes zero.

Of course, teleportation is not possible, so achieving higher movement speed is one of the major goals of technological development. Looking back on the history of railways, the operating speed of the first steam locomotive was about 30 km/h, but it later exceeded 130 km/h for electric locomotives, and now exceeds 300 km/h due to the birth of high-speed railways such as the bullet train. The linear motor car, or maglev trains, can realize speeds exceeding 500 km/h making it the future of railway transportation. The evolution of vehicle technology is also linked to improvements in speed. However, because there are speed limits on public roads, places where cars can demonstrate its performance are extremely limited, but even commercial vehicles can exceed a maximum speed of 300 km/h. In addition, various technologies are also being developed to improve the comfort of movement, such as the maintenance of comfortable air conditioning, the reduction of unpleasant noise and vibration, and the enhancement of interior decoration. All of these contribute to reducing the negative utility when moving, in addition to saving time. When traffic is generally regarded as derived demand, desirable traffic means that travel time, transportation costs, and traffic accidents are reduced to zero.

5.1.2 Academic fields related to transportation

Transportation is closely related to people's daily lives and is integrated with various academic fields to build a unique theoretical system. For example, "transport economics" is a field of economics that analyzes traffic, and "traffic psychology" is a field of applied psychology that considers traffic from the characteristics of human behavior. Alternatively, "transport geography" is a field of geography that captures transport phenomena from regional characteristics. This section mainly focuses on road traffic and explains the basic theories and techniques in traffic planning, traffic engineering, and traffic safety related to the three keywords "cheap", "fast", and "safe".

First, "transportation planning" is an academic field that predicts future urban and environmental conditions and considers necessary transportation developments from various perspectives. It is characterized by thinking about "convenient, cheap and environmentally friendly" transportation in future cities, based on the method of traffic demand forecasting. The related academic fields are diverse, including urban planning, economics, environmental studies, psychology, and engineering. The main focuses include formulating road plans, parking lot plans, public transportation plans, pedestrian and bicycle plans, and comprehensive transportation plans.

"Traffic engineering" is a discipline for safely and smoothly controlling the flow of traffic, mainly for automobiles. It aims to realize "fast and safe traffic without traffic jams" and has features such as describing traffic phenomena by applying physics. It mainly plans the design and operation of road shapes, intersection shapes, traffic lights, and traffic regulations.

"Road safety" is a discipline that aims to deter traffic accidents. Generally, there are three factors that cause a traffic accident: "vehicle", "road", and "human". Among them, safety measures related to "human" are called "3E" because of the acronyms for traffic management/traffic engineering technology (Engineering), the process of ensuring compliance with laws and regulations (Enforcement), and safety education by sociology and pedagogy (Education).

The study of transportation is also a practical study, and by cooperating with each other based on various academic fields, various traffic problems can be solved, and a better transportation society can be developed.

5.2 TRANSPORTATION PLANNING

5.2.1 Traffic survey

The first step in formulating a transportation plan begins with understanding the actual conditions of transportation. By carefully examining what kind of traffic is used when and where, it is possible to consider what kind of traffic improvement is needed in the future. In a traffic volume survey focusing on the behavior of vehicles, the vehicle type, the number of vehicles, speed, changes depending on time, and right/left turn rate at intersections are investigated. In addition, in a person trip survey focusing on the movement of people, attributes, travel time, means of transportation, departure/ arrival points and purpose of travel are investigated. The movement of people is aggregated in units called a "trip", and the traffic volume from the origin to the destination is called the "OD traffic volume".

Survey methods mainly include questionnaire surveys in which residents and users are asked questions and their traffic behavior is extracted, and observation surveys in which the number of vehicles passing through a certain period is counted on the roadside.

In a questionnaire survey, there are revealed preference surveys (RP) that asks questions about past traffic movement while stated preference surveys (SP) asks questions about traffic preferences in specific proposed conditions. The former is used to grasp the actual situation of traffic movement, and the latter is used to forecast the demand for non-existing traffic. In addition, it is difficult to grasp the value of the environment and landscape because they are not directly traded in the market. In such cases, there is a Contingent Valuation Method (CVM) that creates and evaluates a virtual market. By asking the subject to answer questions with Willingness to Pay (WTP) for the value of the landscape, human awareness data can be reflected in the transportation planning.

There are also various types of observation surveys on the roadside, such as data acquisition using various sensors and IC cards, and continuous data collection using GPS (Global Positioning System). There are also traffic volume measuring devices that use fixed sensors such as geomagnetic and tube devices that are installed under the road surface, and image processing devices that use optical flow method and the license plate method to gather data using photos and videos taken from the sky. There are also efforts that are underway to grasp the behavior of vehicles with dynamic sensors. The method of investigating the movement of a vehicle using GPS mounted on a car navigation system is called "probe car investigation", which collects position and speed data by wireless communication. The method of investigating the movement of a person using the GPS of a mobile phone or smartphone is called "probe person survey", and positioning during movement is performed in seconds, and information such as the movement route is accumulated in a database in real time.

In recent years, in addition to GPS data, big data such as mobile phone base station data and Wi-Fi access point data has begun to be used, and it is now possible to grasp the movement of people 24 hours a day, 365 days a year, regardless of the specific means of transportation or location. By using it in combination with the conventional person trip survey, various advantages such as reduction of survey cost and improvement of survey accuracy can be expected. However, since these big data also include personal information such as gender, age, and place of residence, it is necessary to pay attention to confidential processing when using the data from the viewpoint of privacy protection. Basically, data acquisition may require cost and labor, so it is important to select an appropriate combination of survey items and survey methods according to the purpose of utilizing traffic data.

5.2.2 Traffic demand forecast

Since transportation is a derived demand generated from urban activities, it is desirable that transportation is as cheap, fast, and as comfortable as possible. When transportation is regarded as a derived demand in this way, the demand forecast can be constructed under relatively simplified theories such as time saving and cost reduction. Since transportation itself is not the purpose, the model is constructed based on the basic principle of reducing the burden of movement. For example, it is assumed that people select the means of transportation or transportation route that minimizes the monetary value (generalization cost) which are converted from various factors such as the time required to reach the destination, the fare, and the comfort on the move.

The most famous traditional traffic demand forecasting model is the "four-step travel demand model" which was one of the first travel demand models which was developed in the 1950s. In this model, traffic behavior is divided into four parts, trip generation, trip distribution, mode choice, and trip assignment, and demand is calculated by sequentially forecasting each of them.

Step 1: Trip Generation – How many trips are generated?
　　The frequency of origins or destinations of trips in each zone by trip purpose.
Step 2: Trip Distribution – Where do the trips go?
　　OD traffic indicates the traffic volume from Origin to Destination.
Step 3: Mode Choice – What travel mode is used for each trip?
　　Choice of transportation such as walking, car or bus.
Step 4: Route Assignment – What is the route of each trip?
　　Select the route to use.

This model describes actual traffic phenomena using mathematical formulas and is obtained by estimating the relationship with multiple variables related to traffic behavior using statistical methods such as the least squares method. The concept of demand forecast is based on the assumption that the behavioral tendency does not change and assumes that "if many people live there, there is a lot of generated traffic", and "it's easy to get to a nearby and attractive place". It is called an "aggregate model" because it collectively handles traffic demand in a certain area as an analysis unit. The relationship within each step is clear, but it has been pointed out that there is a lack of theoretical consistency between the four steps. However, a certain

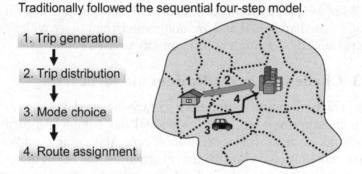

Traditionally followed the sequential four-step model.

1. Trip generation
↓
2. Trip distribution
↓
3. Mode choice
↓
4. Route assignment

Figure 5.1 The four-step travel demand model.

level of estimation accuracy is ensured in the entire analysis area, and it is still used today because of its operability and convenience. It is especially suitable for long-term city planning and forecasting traffic demand in a wide area.

On the other hand, when estimating the impact of a particular project, a more accurate model is needed to reflect individual behavior. A "disaggregate travel demand model" has been developed in which the analysis unit is individual traffic behavior and is not aggregated by area. This model, based on discrete choice analysis, was developed by M. Ben-Akiva and other researchers, and has evolved through various improvements. The early transportation applications of discrete choice models were made for the binary choice of travel mode in 1960s. A feature of this model is that it is assumed that humans make selective behaviors to maximize utility. Here, a prediction model is constructed based on the utility composed of personal attributes and the traffic environment. For example, by comparing the merits (utility) of going by car or bus, it is assumed that a more effective option is likely to be selected stochastically. Since the utility depends on the individual's attributes, enter detailed conditions such as a preference by gender and age, car ownership, and family structure. This makes it possible to analyze the sensitivity to various policies. However, a huge amount of data is required in the case of a wide range of subjects, so it is often used practically for specific policies and issues.

In traffic assignment, it is necessary to determine a rule for which route the road user selects from a plurality of routes leading to the destination. In general, road users prioritize shortening the travel time when selecting a route, so the model is constructed on the assumption that "a person selects the route with the shortest travel time". This is called the "equal travel times principle" advocated by J.G. Wardrop (1952) and is defined as follows. The journey times on all routes used are equal and are not greater than those which would be experienced by a single vehicle on any unused route.

5.2.3 Challenges in traffic demand forecasting

In the traffic demand estimation so far, the model is constructed on the assumption that "each individual has complete information on options such as transportation means and routes, and people always make rational choices". In other words, the result of acting to maximize the utility of the individual is calculated as the estimated demand value.

However, in the real world, not all information is available, and people do not always make rational decisions. When going out, it is often the case that traffic behavior is decided immediately without obtaining traffic information. In such cases, there is a certain bias in rational judgment because the behavior is decided based on experience and intuition so far. This is called a heuristic bias in psychology. A. Tversky and D. Kahneman (1974) found that there are three heuristics in judgment under uncertain circumstances which is representativeness, availability, and adjustment and anchoring, each of which is biased. Considering each bias by substituting for traffic behavior, it becomes as follows.

- Representativeness bias: overestimate the probability of something that seems to be representative of a category. A concept similar to stereotypes. If there is a fixed concept that transferring public transportation is a burden in terms of time and comfort, it is difficult to increase the number of users even if a convenient transfer system is constructed.
- Availability bias: things that are easy to recall are prioritized and evaluated. It is a mental shortcut that relies on immediate matters that come to a given person's mind. For example, a person who travels by car every day tends to leave the door with their car keys without thinking when going out.
- Adjustment and anchoring: overemphasis on the characteristics of the initial information. This is a phenomenon in which an individual makes an initial idea and response based on one point of information and makes changes driven by that starting point. If the first bus you board is delayed by 15 minutes, that information will become an anchor and the evaluation of the bus will not be judged correctly.

It has also been confirmed that even if there is something better than the current means of transportation and routes, people tend to continue their traffic behavior. It is common to see the use of familiar paths daily, and this is called "status quo bias".

Such biases or statistical fluctuations are treated as probabilistic estimation errors in traffic demand forecast, and do not pose a problem in situations where the proportion does not increase. This is because the target traffic is only derived demand, and when many samples are aggregated and processed as the traffic volume, the bias due to individual differences becomes relatively small.

5.2.4 Comprehensive transportation system

There are various means of transportation in the city such as walking, bicycles, cars, buses, and railways. It is extremely important in transportation planning not only to consider the optimization of individual transportation, but also to identify the advantages and disadvantages of the various means of transportation, and to consider the role and cooperation methods of each transportation.

Traffic demand fluctuates greatly depending on the time of day, day of the week, and season. Introducing transportation with high capacity to meet peak demand will be wasted during off-peak hours. Therefore, it is essential to develop an appropriate transportation system in consideration of the moving distance and the number of users. For public transportation, there are mass transit such as railways and subways, trams and buses for medium-duty transportation, and taxis for small-volume individual movement. In individual transportation, the movement density of bicycles and walking is high, but the moving distance is short. Automobiles can cover short to long distances, but fluctuations in demand often cause road congestion due to their low movement density.

When various means of transportation is divided into inter-city, intra-city, and intra-district movement depending on the role of each

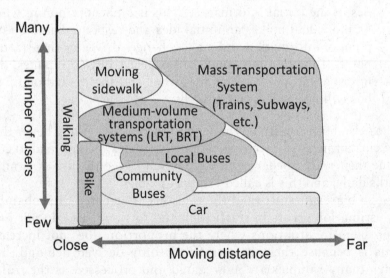

Figure 5.2 Relationship between moving distance and number of users by means of transportation.

means of transportation, the hierarchy of the transportation network can be expressed as follows. High-speed transportation plays a role between cities, and within cities, appropriate transportation is used depending on the transportation density. The most familiar transportation in the neighborhood is basically walking, and mainly low-speed movement under 30km/h. If the hierarchy of the road network breaks down, the passing traffic will penetrate into the inner city, congesting not only the living roads, but also making the traffic environment extremely dangerous. In the same way, if the hierarchy of the public transportation network is disrupted, express and punctuality will be hampered, transfers will be inconvenient, and the network will be extremely confusing for users. In addition, it is ideal that the road network and public transport network are well-balanced in the city. In recent years, the public transportation network has been drastically reduced due to the increasing use of automobiles, and the balance is severely skewed. In order to cope with environmental problems and an aging society, it is desirable to establish an appropriate relationship between the two networks.

To build a comprehensive transportation system, it is necessary to formulate an "Urban Transport Master Plan" for ideal future transportation based on analysis of current traffic conditions and forecast of future traffic volume. On the premise of a comprehensive plan that includes land use and transportation, the Urban Transportation Master Plan should formulate detailed plans such as routes and

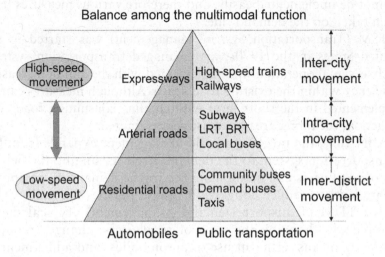

Figure 5.3 Hierarchical nature of urban transportation.

nodes of roads and public transportation, as well as measures and programs to promote the plans. Relevant plans include the following.

- Metropolitan structure and land use planning: reflection of regional master plan and city master plan
- Road planning: improvement plans for city planning roads, etc.
- Public transportation plan: development plan for subway, new transportation system, LRT, BRT, bus, etc.
- Transport hub plan: plan for station square, bus terminal, etc.
- Others: park and ride, staggered commuting, enlightenment campaign, etc.

5.2.5 Traffic management

Many traffic problems, such as traffic congestion, are caused by an imbalance between the traffic demand and the transportation facility supply. This is partly because the development of transportation facilities such as roads and railways cannot keep up with the increased traffic demand due to rapid urbanization and population growth. To maintain an appropriate balance between traffic demand and supply, it is necessary to either take measures to increase the supply of transportation facilities or to curb the traffic demand itself. At first, the improvement of facilities should be considered to balance the traffic supply and demand for the purpose of continuous growth of the city. In many cases, road construction cannot be implemented easily, and therefore various measures have been taken on the demand side.

TSM (Transportation System Management) was started in the United States in the 1970s, and managed transportation systems such as signal operation was implemented with the aim of increasing efficiency within the existing road space. Although measures can be implemented in the short term without time consuming road construction, the target area of adjustment is limited.

After that, the target of traffic management expands from the transportation system to the demand itself. Changing the behavior of road users can manipulate the traffic demand to an appropriate level to reduce congestion. These policies are collectively called TDM (Transport Demand Management). Typical methods include changing the time of road users, changing routes, changing means, efficient use of automobiles, and adjustment of trip generation. In addition to balancing transportation facilities and transportation demand, the aim is to keep the demand itself within the appropriate capacity and to prevent excessive facility development.

The subject of traffic management refers to the psychology of the person who causes the traffic behavior. MM (Mobility Management) is defined as a "communication-centered measure that encourages the voluntary change of individual and organizational mobility in the desired direction for both society and individuals". By applying social psychology and individually calling for traffic behavior that considers the environment and health, MM has achieved results in alleviating traffic congestion and promoting the use of public transportation.

5.3 TRANSPORTATION ENGINEERING

5.3.1 Understanding traffic phenomena

Traffic is a tangible movement mediated by human will, and like the flow of water and air, the movement of pedestrians and automobiles can be regarded as a flow on the road from a macroscopic perspective. The number of vehicles that have passed a certain point during the measurement time is called the traffic volume, and the measured traffic volume converted per unit time is called the traffic flow rate. The traffic density indicates "the number of vehicles existing per unit distance of the road at a certain time", and the space mean speed is "the average speed of vehicles existing per unit distance on the road at a certain time". It is known that the following conservation law of flow rate holds between this traffic flow rate, traffic density and space mean speed.

Traffic flow rate Q = Traffic density K × Space mean speed V

The relationship between the traffic flow rate Q and the space mean speed V is called "Q–V curve" and is as shown in Figure 5.4. From this figure, there is a maximum value for the traffic flow rate, which is evident in the maximum highway traffic capacity. Furthermore, even if the traffic flow rate is the same, there are two speed states with the critical speed as the boundary, the upper part showing the free flow and the lower part showing the congested flow.

Traffic congestion is a phenomenon that occurs when traffic demand exceeds traffic capacity. Traffic congestion is caused by a Bottleneck, which has a relatively low traffic capacity compared to

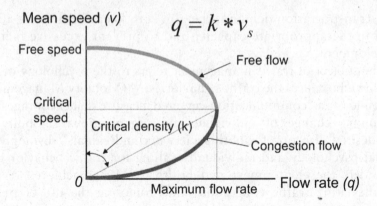

Figure 5.4 Relationship between traffic flow rate, density, and space mean speed.

the upstream section. Traffic congestion is the queue at the convoy length that occurs in the upstream section of the bottleneck.

Figure 5.5 shows the changes in traffic flow rate and queue length over time from the occurrence of congestion to its elimination. The queue length appears when the traffic demand exceeds the traffic capacity, and the queue length becomes maximum when the excess

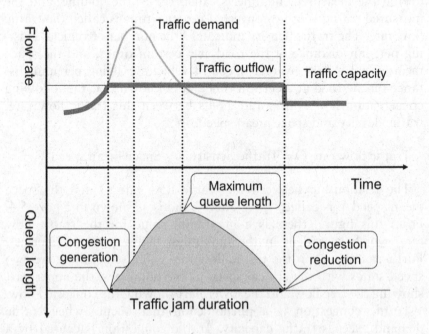

Figure 5.5 Time variation of traffic flow rate and queue length.

traffic demand is resolved. After that, the traffic jam will continue, and the traffic jam will disappear only after all the vehicles accumulated during the excess demand have passed through the bottleneck. It should be noted that the time when the traffic demand exceeds the traffic capacity is different from the time when the maximum queue length appears and the duration of the congestion.

5.3.2 Traffic simulation

With the progress of traffic engineering, simulation techniques have been used to solve road traffic problems. This is because the actual traffic phenomena are often large-scale and complicated, and the components of the transportation system include uncertainties that fluctuate stochastically.

Traffic simulation can be roughly divided into a macro model that treats the flow of traffic as a continuous fluid and a micro model that duplicates the behavior of individual vehicles in detail. The former is called "macro traffic simulation" and is suitable for reproducing the flow of traffic over a wide area of the entire city or for long-term predictions. The latter is called "micro-traffic simulation" and ever-changing traffic conditions can be represented by processing the behavior of each vehicle in a relatively small area. In recent years, as computers have become faster, micro-traffic simulations have also begun to be used in a wider range of analysis.

Various traffic simulations have been developed in many countries so far, and their characteristics are different for each model. Generally, the following input data is required.

- Data on road networks (traffic volume, link capacity, link length, number of lanes, etc.)
- Regulation/control data (no right/left turn, one-way street, signal control parameters, etc.)
- Data on traffic demand (intersection branching fraction, OD traffic volume, etc.)

Traffic volume, route travel time, traffic jam length, etc. are outputs found by executing traffic simulation, and environmental loads such as CO_2 and NOx, energy consumption, etc. are calculated based on these values. For example, in a micro traffic simulation, the behavior of a vehicle is reproduced in units of 1 second, and the amount of CO_2 emitted within the next 1 second is calculated from the speed,

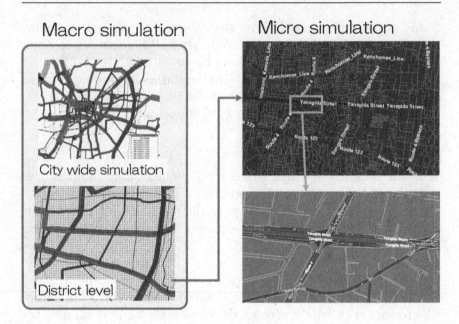

Figure 5.6 Macro and micro traffic simulation.

acceleration, vehicle type, etc. of a certain vehicle. As a result, the environmental load is calculated by totaling the values for all vehicles. By analyzing such big data, it is possible to predict when, where, and how much traffic congestion will occur with high accuracy, and the analysis results are used in various situations from road improvement to traffic assessment.

5.3.3 Congestion countermeasures

Congestion occurs when traffic demand exceeds traffic capacity, so traffic congestion countermeasures can be broadly divided into either temporally or spatially distributing traffic demand or increasing traffic capacity so that it does not exceed the traffic capacity. Since the former traffic management has already been explained, the latter method of increasing the traffic capacity will be outlined here.

First, traffic congestion countermeasures begin by identifying bottlenecks in traffic capacity. After that, the traffic capacity and traffic demand must be found, and the magnitude of excess traffic demand and the cause of its occurrence must also be investigated. Most of the congestion points on general roads can be seen at intersections. Signal adjustment is relatively easy to implement as a countermeasure against traffic congestion at intersections.

Appropriate traffic signal control includes adjustment of blue time split and reduction of signal phase numbers. Alternatively, if the road width is relatively large, the traffic capacity will be increased by improving the road markings. Specifically, the width per lane can be reduced to increase the number of lanes, and a dedicated right/left turn lane can be newly established, extended, or expanded. If such measures cannot be implemented, large-scale intersection improvement will be required. The width of the road must be widened by the setback of land along the road, and then the number of lanes can be increased. If the excess traffic demand still cannot be resolved, it will be necessary to create a three-dimensional intersection or to develop a bypass road that detour the intersection.

Traffic congestion also occurs at the entrance of parking lots to large-scale facilities. This is due to the slowdown of cars near the entrance, congestion in the facility, and people waiting for open parking lots. Methods for improving this includes the installation of deceleration lanes, the securing of lead-in paths that are not disturbed by the parking behavior in the facility, and the expansion of parking lots. In addition, congestion at the confluence of expressways can be reduced by improving lane operation, and congestion at toll gates can be reduced by introducing the ETC (Electronic Toll Collection System).

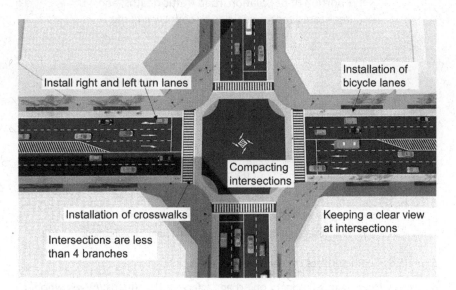

Figure 5.7 Examples of traffic congestion countermeasures.

Some traffic jams, such as chronic traffic jams, occur constantly, while others occur non-steadily or suddenly due to construction work, traffic accidents, or disasters. Flexible operation of lanes and shoulders is required to deal with such fluctuating traffic environments. It is also desirable to prepare for future uncertainties by giving surplus to the shoulders at the design stage.

5.4 TRAFFIC SAFETY

5.4.1 Situation and cause of traffic accident

Every year, about 1.35 million people are killed in road accidents with the number of injured reaching 50 million. This makes it the eighth leading cause of death for all ages and is the number one cause of death for children and adolescents between the ages of 5 and 29, making traffic accidents a profoundly serious problem. According to a WHO survey, the number of deaths per 100,000 population decreased slightly from 18.8 to 18.2 between 2000 and 2016, but the total number of deaths per year increased from 1.15 million to 1.35 million, indicating that the total number of casualties is growing worldwide. Road fatalities in low-income countries accounts for 13% of the world's total while only

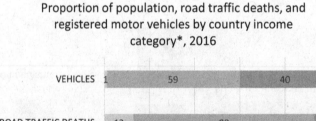

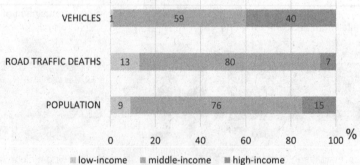

Figure 5.8 Characteristics of road traffic fatalities.

Source: Global status report on road safety 2018 (https://www.who.int/ violence_injury_prevention/road_safety_status/2018/en/).

accounting for 1% of the world's vehicles. In comparison, high-income countries account for 40% of vehicles, but the death toll only accounts for 7%. This indicates the dangerous gap in traffic safety in developing countries.

The causes of traffic accidents are classified using the 4M analysis, which is used for the analysis of the causes of accidents and disasters and the examination of countermeasures.

1. Man: caused by drivers and pedestrians
2. Machine: caused by vehicles, traffic lights, etc.
3. Media: due to road environment and traffic environment
4. Management: due to the organization, management and education system

Among them, the first 3M (Man, Machine, Media) factors are easily associated with traffic accidents, and are tabulated as accident factors in traffic accident statistics of major countries such as the United States, Germany, France, and Japan. Although the causes of accidents differ depending on the region, human factors are the most common cause of traffic accidents, and they often account for more than 90% of all accidents. Therefore, the cause of traffic accidents is said to be human error. According to Japan's traffic accident statistics, the most common human factor in accidents is the delay in discovery such as carelessness and unconfirmed safety, which accounts for more than 70% of the human factors. On the other hand, due to the improvement of vehicle performance, vehicle factors have decreased from 7.2% (1953) to 0.2% (2019), and environmental factors such as poor visibility and poor road surface conditions have remained at a low level of 3.6% (2019).

5.4.2 Analysis of traffic accidents

There are many factors involved in the background of a single road accident. For example, a certain rear-end collision was directly caused by the deceleration of the vehicle in front and the delay in the driver's perception leading to insufficient braking. However, the tires of the car were worn, and the road surface was wet with rain, which increased the braking distance and caused an accident. As such, road accidents are often the result of a combination of different factors. Therefore, when investigating the cause of a traffic accident, it is necessary to investigate from various

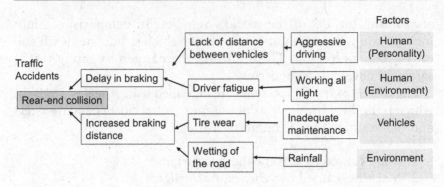

Figure 5.9 Causes and chains of traffic accidents.

perspectives. In addition, there are also indirect causes that are linked to direct causes, so when taking drastic measures, the relationship between them should be fully understood. In the above example, the root cause of the accident could be the driver's safety awareness (or working environment), inadequate vehicle inspection, and the natural environment.

VTA (Variation Tree Analysis) is a method for retroactively investigating the cause of an accident. This is a method of considering the chain of causes of an accident by branching events related to the accident in a tree shape and assuming unusual actions and judgments in the order of occurrence.

Each accident has its own cause, so detailed accident countermeasures should be taken case-by-case. On the other hand, it is also important to classify similar accidents and take comprehensive measures. For that purpose, it is necessary to collect the data of traffic accidents in a unified format (accident original form) and accumulates and manage it as traffic accident statistics.

The collected data is statistically processed and the number of accidents, fatal accidents, casualties, fatalities, etc. is published regularly. The data can then be used to grasp the secular change of traffic accidents and measure the impact before and after countermeasures. In addition, standardized indicators are used to make relative comparisons of traffic accident situations in different regions. Taking the number of accidents as an example, the unit of the number of accidents divided by the population (cases/person) and the unit of the number of accidents divided by the road extension (cases/km) can be used. Alternatively, there is also a unit divided by the number of vehicles owned (cases/vehicle) and a unit divided by the vehicle mileage (cases/vehicle km). Accident countermeasures can be implemented efficiently by objectively comparing and extracting locations and areas where traffic accidents occur frequently.

In addition, the probability of a traffic accident is extremely low but is always changing, so statistical verification is required to measure the effectiveness of traffic safety measures. For example, suppose that traffic safety measures are implemented at an intersection where 15 accidents occur each year, and the number decreases to 12 the following year. The number of accidents has decreased in one year, but it is necessary to judge whether this is a countermeasure effect or whether there were fewer accidents by chance this year. In general, assuming that the accident intervals are completely random, determining whether the values obtained are within the range of accidental fluctuations or significantly different from the theoretical distribution (Poisson distribution) is necessary. The significance level used here indicates the probability of misjudgment and is usually set to 5% or 10%, and if it exceeds the theoretical number of occurrences, it is judged that there is a statistical change.

Meanwhile, the occurrence of traffic accidents depends on the day of the week, the season, and economic activity. Observing at the fluctuations in traffic accidents over the year, the number of accidents increases when economic activity is strong, and the number of accidents decreases when economic activity is stagnant. In this way, traffic accidents include seasonal fluctuations and periodic fluctuations, so in order to evaluate the effects of countermeasures, it is necessary to make adjustments such as setting comparative targets for one to several years before and after the countermeasures.

5.4.3 Traffic safety measures

Since the causes of traffic accidents are diverse, traffic safety measures are also diverse. Road safety activities are commonly referred to as the 3E measures. 3E is an acronym for the three areas, namely, education, engineering, and enforcement. It is said that it started in the United States in the 1920s when various traffic problems became apparent due to the spread of automobiles.

In the field of education, there is traffic safety education from early childhood, dissemination of safety equipment such as seat belts and helmets, traffic safety campaigns by companies and local communities, and training when renewing licenses. In the field of engineering,

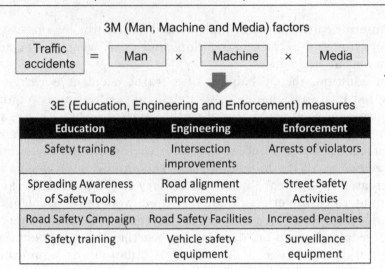

Figure 5.10 Causes and countermeasures for traffic accidents.

there is improved road alignment, improved pavement, improved intersections, and the maintenance of road safety facilities such as traffic lights and road signs. Vehicle safety devices such as airbags and vehicle ABS also play a major role. Traffic enforcement is also necessary to reduce drivers who drive dangerously and make them comply with traffic rules to prevent traffic accidents. Countermeasures include arresting offenders, police patrols, increasing penalties, and installing surveillance equipment.

Other road safety measures also include emergency lifesaving, post-accident treatment, rehabilitation, and victim care. From the viewpoint of preventing traffic accidents, it is important to create a safe and secure city in city planning by promoting the use of public transportation and designing blocks to prevent passing traffic from entering the living space. There are various research fields related to traffic safety, such as sociology, pedagogy, engineering, physics, statistics, psychology, law, and economics. In that sense, road safety is comprehensive and interdisciplinary cooperation is extremely important.

There has been a shift in the basic approach to traffic safety from an individual field to a comprehensive field. Instead of focusing on the parties directly involved in traffic accidents, it has gradually become recognized as an issue for the entire transport sector. The impetus for this was the "Vision Zero" passed by the

Swedish parliament in 1997. This is based on the long-term goal of reducing the number of deaths and serious injuries caused by traffic accidents to zero. However, it is also essentially based on the establishment of a theory of responsibility that places the cause of accidents on the entire transportation system and a philosophy that no longer tolerates a transportation system that exists at the expense of human life. This idea spread to neighboring countries, then to the United Kingdom, Australia, and the United States, and was later positioned as one of the 17 global goals in the Sustainable Development Goals (SDGs) adopted by the United Nations General Assembly in 2015. In other words, achieving comprehensive road safety in the context of sustainable urban development is a major challenge for the future.

REFERENCES

Moshe Ben-Akiva, Steven R. Lerman. 1985. *Discrete Choice Analysis: Theory and Application to Travel Demand*, Cambridge, MA: The MIT Press.

John A. Groeger. 2011. How Many E's in Road Safety? In *Handbook of Traffic Psychology*, ed. B.E Porter, 3–12. London: Academic Press.

Institute for Traffic Accident Research and Data Analysis. 2019. *Statistical Data*. https://www.itarda.or.jp/ (accessed August 10, 2020).

Japan Society of Traffic Engineers. 1991. *Key Points for Intersection Improvement*, Tokyo: Maruzen Publishing Co., Ltd. (in Japanese).

Japan Society of Traffic Engineers (ed.). 2000. *Simulation for Traffic Engineering Made Simple*, Tokyo: Japan Society of Traffic Engineers (in Japanese).

Japan Society of Traffic Engineers. 2018. *Road Traffic Technology Handbook 2018*, Tokyo: Maruzen Publishing Co., Ltd. (in Japanese).

Hiroshi Konno (ed.). 1981. *New Edition Urban Planning*, Tokyo: Morikita Publishing Co., Ltd. (in Japanese).

Hisashi Kubota, Takashi Ohguchi, Katsumi Takahashi. 2010. *Read and Learn Traffic Engineering and Planning*, Tokyo: Rikoh Tosho Co., Ltd. (in Japanese).

Tsuneo Matsuura. 2014. *Hints for Traffic Accident Prevention from Statistical Data*, Tokyo: Tokyo Horei Publishing Co., Ltd. (in Japanese).

Michael D. Meyer, Eric J. Miller. 2001. *Urban Transportation Planning: A Decision-Oriented Approach*, 2nd ed., Boston: The McGraw-Hill Companies, Inc.

Norbert Oppenheim. 1995. *Urban Travel Demand Modeling: From Individual Choices to General Equilibrium*, New York, NY: John Wiley & Sons, Inc.

Ayako Taniguchi, Haruna Suzuki, Satoshi Fujii. 2007. Mobility Management in Japan: Its Development and Meta-Analysis of Travel Feedback Programs, *Transportation Research Record: Journal of the Transportation Research Board*, 2021(1):100–109.

Claes Tingvall, Narelle Haworth. September 6–7, 1999. Vision Zero – An Ethical Approach to Safety and Mobility, In 6th ITE International Conference Road Safety & Traffic Enforcement. Melbourne.

Amos Tversky, Daniel Kahneman. 1974. Judgment under Uncertainty: Heuristics and Biases, *Science*, 185(4157):1124–1131.

John G. Wardrop. 1952. Some Theoretical Aspects of Road Traffic Research, *Proceedings of the Institute of Civil Engineers*, 2:325 Norbert Oppenheim 378.

World Health Organization (WHO). 2018. *Global Status Report on Road Safety 2018*. 24. https://www.who.int/publications/i/item/9789241565684

Hirotaka Yamauchi, Kenzo Takeuchi. 2002. *Transport Economics*, Tokyo: Yuhikaku ARMA (in Japanese).

Chapter 6

Consider cities based on the transportation

6.1 THE ESSENCE OF TRANSPORTATION

6.1.1 Movement from the perspective of psychology

Transportation is broadly defined as "the spatial movement of people, goods, and information by the will of the person". Transportation may include information in a broad sense, but the movement of information is treated separately as communication, and transportation generally refers to the movement of tangible objects such as people and things. In this section, the essential meaning of human movement is described.

Why do people move? There is a lot of research on the motives of movement from the past. In the field of psychology, Abraham Maslow (1943, 1954) categorized human needs into five levels, assuming that human needs grow toward realization. The most basic is the instinctive or a physiological desire to sustain life, followed by the desire for safety so that one can lead a stable life. When both are satisfied, humans have a social desire to be recognized in society. When they play a role in society, they need higher levels of socially respected approval, such as status and honor. Ultimately, we want self-actualization to be the ideal self we imagine, rather than comparing it to others.

Reichman (1976) categorized mobility into three categories based on the need for transportation behavior: mandatory mobility, maintenance of livelihood, and discretionary mobility. In contrast to the transportation purposes (commuting to work, school, business, private affairs, and returning home) in the Person Trip Study to understand human transportation behavior, they can be organized as follows.

- Mandatory: commuting, going to school, returning home
- Maintenance: shopping, going to hospital, private movement
- Discretionary: social activities, recreation, entertainment

Mokhtarian et al. (2015) organized the travel motivations resulting from Maslow's five requirements to suit three transportation objectives and pointed out that the lower three of Maslow's layer account for most of the transportation objectives.

1. Self-actualization: travel for curiosity, aesthetic appreciation, experimenting with new modes or routes
2. Esteem: travel for status, preferring modes perceived to be higher status
3. Social needs/love and belonging: travel for social activities, volunteer/club/religious activities
4. Safety needs: travel for work, medical, exercise
5. Physiological needs: travel for grocery shopping, eating out

Thus, most traffic behaviors are obligatory or indispensable for maintaining a living. On the other hand, there are also traffic behaviors that can be freely decided within the discretion of the individual.

As a theory of human motivation, there is the theory of self-determination that explores the path to intrinsic motivation. In this theory, motivation of movement can be divided into the following three: intrinsic things that are inherent in people, extrinsic things that are external, and motivation things that are not recognized by individuals. The concept of utility in economics includes both intrinsic and extrinsic motives, but in transportation, the main focus is on extrinsic motives. According to Mokhtarian et al., people's intrinsic motivations include autotelic, hedonic, and experiential, with hedonic motivations appearing in the short term, whereas they become eudaimonic when they are long-term. Meanwhile, extrinsic motives include instrumental, utilitarian, and functional.

Elucidating people's motivations for moving is the starting point for considering the nature of transportation. When transportation is considered as a derived demand and obligatory travel accounts for the majority of urban transportation, the essential task of transportation planning is to deal with urban activities, which are the primary demand. However, as leisure time increases and discretionary travel increases, there will be more cases where transportation behavior itself becomes an objective and a primary demand. As commuting and shopping, which used to be derived demand, change to teleworking and online shopping, the role of transportation, which is the primary demand, will become increasingly important in the future.

6.1.2 Movement from the perspective of economics

In economics, which attempts to theoretically explain the behavior of various people and organizations exchanging goods (goods and services) and money in the marketplace, transportation can be regarded as a service, including the case of traveling by oneself. Transportation is generally a derived demand because it occurs to carry out some other economic activity, and services are provided in tandem with specific points such as origins and destinations. It has the characteristic of time consumption because it takes time to move, and the characteristic of self-sufficiency because it can move by itself.

In the "Principles of Economics", economist Alfred Marshall of the early 20th century analyzed derived demand, taking architecture as an example, and described that all intermediate inputs including necessary labor is derived demand. However, transportation may become a primary demand such as walking and driving, and the characteristics as derived demand are relative. In this way, transportation is a derived demand goods with many exceptions compared to goods and services that seem to have no use other than intermediate inputs.

Economics analyzes the relationship between price, supply, and demand, assuming that economic agents behave rationally within given constraints. In other words, the price of transportation services is determined by the relative relationship between supply and demand. According to Marshall, derived demand is inelastic in the following four cases, as opposed to the primary demand of the final product. Elasticity indicates the ratio (A/B) of the rate of change A of one variable to the rate of change B of another variable. For example, if the rate of decrease in demand is −5%, which is less than the rate of increase in price of 10%, the "price elasticity of demand" is −0.5, which is less than 1 and is said to be inelastic. In the following four cases, the impact of the decrease in transportation on the fares is examined, assuming that the final product is a land use (activity) and the factor of production is transportation. In both cases, the rate of increase in fares is higher than the rate of decrease in traffic demand.

1. When the production factor is a basic good of the final product
   Insufficient transportation services in the absence of alternative transportation services can result in relatively significant increases in transportation costs. For example, in a society that relies heavily on automobiles for urban activities, a decrease in

the supply of automobiles will significantly increase transportation costs.

2. When the demand curve for the final product is inelastic

Even if there is a shortage of transportation services and the activity price rises, if the decrease in activity demand is small (inelastic), the price of the shortage of transportation services can rise significantly. For example, if bus transportation capacity declines and admission to tourist destinations rises, but demand for tourist destinations does not decrease significantly, bus fares can rise significantly.

3. When the ratio of spending on production factors to the total cost of the final product is small

If the cost of a transportation service accounts for a small percentage of the total cost, the price of the transportation service tends to rise. For example, if the cost of transportation to the theater is low compared to the very high cost of the theater, it is easy to accept the increase in transportation costs.

4. When the supply curve of other factors of production is inelastic

When using multiple transportation services, if the supply of other transportation services cannot be increased in line with the rate of increase in fares, the decrease in transportation services can lead to a significant increase in fares. For example, when commuting by connecting between a bus and a railway, the bus fare will increase if the railway operation level does not change even if the number of buses is reduced.

It is generally known that the elasticity of demand becomes elastic with respect to prices for goods with a lot of luxury goods and competitive goods and becomes inelastic with respect to prices with goods with few necessities and competitive goods. As for the transportation, the former is movement at the discretion such as sightseeing in places with various modes, and the latter is movement for obligatory or livelihood maintenance in places with few modes of transportation. In this way, the content of transportation services is diverse, and changes in supply and prices throughout the market also affect other transportation services and urban activities, which are the primary demand.

In addition, time factors are important for understanding the equilibrium between supply and demand. According to Marshall, prices depend on supply, demand, and time factors, and in addition to the current market prices, it can be classified into following categories:

short-term normal price for several months to one year, long-term normal price for several years, and even more ultra-long-term normal price. In general, shorter times have a greater impact on the price of demand, and longer times have a greater impact on production costs. This is because it takes time to adjust production volume, such as capital investment.

Considering the relationship between transportation and land use, transportation facilities can be regarded as supply for responding to land use linked to demand. In the process of balancing traffic and land use in the market, the increase in traffic capacity is gradual on the time axis and will change in the long run under the influence of changes in social conditions.

6.1.3 Transportation as a primary demand

So far, most of the traffic has been described as derivative demand. For example, when commuting by bus, working is considered the primary demand, and going to work is considered derived demand driven by primary demand. Getting on the bus is a way to achieve the original purpose. Meanwhile, for some leisure and tourism activities, transportation itself may be the purpose. Traffic behavior itself will be the primary demand when driving along the coastline or looking at beautiful scenery through the windows of a tourist train. Transportation is also a primary demand when children want to fly or take high-speed trains. In transport economics, derived demand is defined as "demand generated to achieve other purposes such as going to work or school" and primary demand is defined as "demand aimed at consuming the transportation service itself".

The primary and derived demands for transportation are conceptually understandable, but they are difficult to measure quantitatively. This is because primary and derived demand fluctuates depending on individual tastes, time, and place. For example, if a person who likes driving drives a car as a hobby, that is the primary demand, but when fatigue accumulates after a certain period of time, he wants to arrive at the destination early and gradually changes to derived demand. In addition, the scenic coastline may be the primary demand, but the inside of a tunnel, it will be a derived demand, and it often fluctuates during movement.

Here, the quantification of primary and derived demand is based on the utility acquired by the user during a trip from origin to destination for a certain purpose. By accumulating the various events

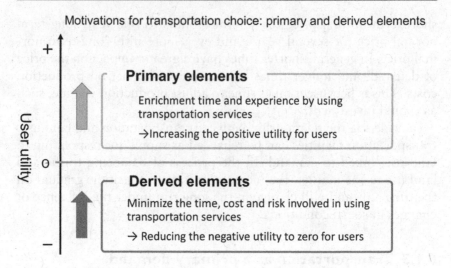

Figure 6.1 Primary demand and derived demand for transportation.

that occur during the trip, positive utility can be defined as primary demand, and negative utility is defined as derived demand. In other words, if the use of the transportation service feels like a fulfilling time or experience for oneself, it is assumed that the user has received a positive utility. Travel is considered a negative utility if users want to minimize the total amount of time, effort, cost, risk, etc. when using transportation services. If the utility of both can be measured, it is possible to distinguish between the primary demand and the derived demand depending on which of the utilities is larger.

6.2 TRANSPORT-BASED CITY PLANNING

6.2.1 What is the transport-based city planning?

When considering transportation from a city planning perspective, transportation policies target derived demand with the main purpose of providing "cheap, fast and safe" transportation. Meanwhile, if transportation is widely regarded as containing primary demand, it is reaffirmed that transportation improvement is a part of community development. In other words, it is possible to create a more attractive urban space by actively capturing the joy of movement as part of the transportation function and reconstructing both the moving space and the activity space. Such movements have been active all over the world since the 1970s, when the expansion of commercial facilities in the suburbs and the decline of downtown areas became apparent due to the progress of motorization. It is also an attempt to review an excessively automobile dependent society and return it to an environment-friendly society. Regulations on the inflow of cars into the city center, transit malls that prioritize public transportation and pedestrians, etc. are all policies aimed at revitalizing the city center. By intentionally limiting the convenience of cars, the bustle of the city can be regained intentionally.

Since the latter half of the 1990s, efforts to revitalize cities using transportation measures have been called "transport-based community development" in Japan. This is also a transportation planning that contributes to value-creating town development that builds a comfortable town through the realization of a desirable life image. In this section, the subject is expanded from community development to city planning in a broader sense, and "transport-based city planning" is explained. In conventional transportation planning, land use is regarded as a given condition, and a society that develops through economic growth is assumed. After that, future population estimation and future demand forecasting is carried out, and various transportation plans have been formulated based on the forecast results. In other words, we consider supply from demand. However, there were also issues such as forecast uncertainty and the adequacy of the response to the forecast. Therefore, a comprehensive approach is proposed by using the idea of reversal to create a flow of thinking about demand from supply and incorporating it into a traditional planning theory.

6.2.2 The method of transport-based city planning

Even if urban development is considered from a transportation perspective, the conventional planning theory should not be ignored. Rather, the first step is to think of ways to integrate it with traditional planning theory. There are three important perspectives on the transport-based city planning. It is a desirable vision of the future, coordination with the various existing plans of the city, and the ability to execute the plan smoothly.

1. Shareable future vision: transport-based city planning is a part of city planning, and most importantly, discussing what the city should be in the future and drawing a vision for the

future is a crucial step. The created future vision should be an ideal image shared with the citizens, and it will take many years to realize it. If it is related to the entire city, it should be positioned in the comprehensive plan or city planning master plan, which is a legal administrative plan, and it is important to review it in consultation with citizens, governments, experts, etc. When creating a vision for a narrow area, the community residents and the local shop owners should be proactively involved in the planning process.

2. Coordination between plans: there are numerous plans within a city. It is important to coordinate between these plans, from high-level plans to low-level plans, from comprehensive plans to individual implementation plans. In addition, looking at the target fields related to city planning, there are various plans such as central city area revitalization plans, vacant house restoration plans, road improvement plans, and industrial promotion plans. Different fields have different management entities, and in some cases, there is a trade-off relationship. For example, there are departments that want to attract large-scale commercial facilities to the suburbs for industrial promotion, and departments that want to suppress them to revitalize the central city area. Urban transportation strategies need to be planned, prioritized, and managed over time, paying attention to the interrelationships between these plans.

3. Plan feasibility: pursuing the ideal image may increase the risk that the plan will turn out to have been nothing but a pie in the sky. Especially when pursuing an ideal image that is far from reality, it can cost a lot of money and time. If there are no signs of realization in the plan, the motivation of stakeholders may be reduced and the plan itself may disappear. To avoid this, a strategy that combines a highly feasible plan and a less feasible plan can be considered. It is important to prepare the planning steps so that you can climb the stairs little by little and create a roadmap that gradually approaches the target image. In addition, it is also necessary to have the idea of "implementing from where possible", such as implementing in order from the place with the highest profitability, or implementing in order from the place where agreement with residents and related parties can be reached.

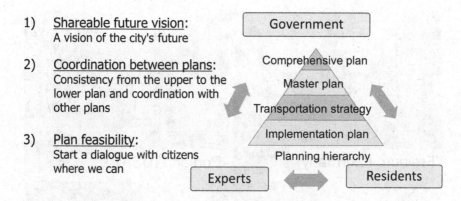

1) <u>Shareable future vision</u>:
   A vision of the city's future

2) <u>Coordination between plans</u>:
   Consistency from the upper to the lower plan and coordination with other plans

3) <u>Plan feasibility</u>:
   Start a dialogue with citizens where we can

Figure 6.2 Three perspectives of transport-based city planning.

6.2.3 Example of transport-based city planning

There are many cases around the world where cities have been regenerated by improving traffic. Freiburg in Germany is one of the cities that have been aiming to reduce car traffic since the 1970s. The city center is open to pedestrians as a transit mall, with tram stations within 500 meters of most parts of the city. At the same time, Portland in the United States also changed their policy from an automobile to a public transport focused city, adopting many transport policies including transit malls in the city center. Because of these policies, it was selected as the most livable city in the United States. Since the 1970s, Curitiba in Brazil has completed the development of a public transportation network centered on pedestrian roads and bus transportation in the city center. In particular, it is characterized by setting five urban axes and building a centralized urban structure through the corridor – type development that conforms to bus rapid transit (BRT).

In the 1980s, the review of public transportation became active in France as well. Particularly in Strasbourg, the new tram was completed in 1994 and urban design combine to create a good example of a city that is fun to walk around.

In the 1990s, many cases of transport-based city planning began to be seen in Japan. Here, the process will be explained for Utsunomiya City, which has a population of 510,000 and is one of Japan's regional hub cities. The plan for the new transportation system in Utsunomiya City began in 1993, and its main purpose was to eliminate road congestion. Around 2000, it was discussed as a means to revitalize the city center against the backdrop of the decline of the central city area. The comprehensive plan and city planning master plan was formulated in

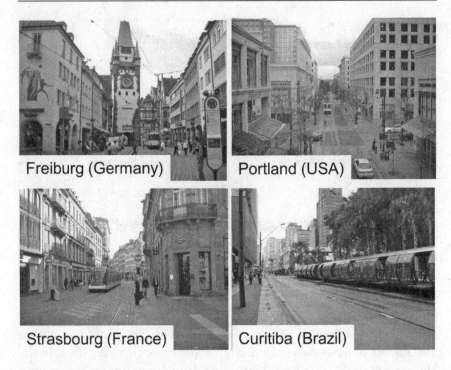

Freiburg (Germany)   Portland (USA)

Strasbourg (France)   Curitiba (Brazil)

Figure 6.3 Example of transport-based city planning.

2008 and was comprehensively positioned as a response to the declining population. A network-type compact city was decided as the urban structure to be aimed at in Utsunomiya, and light rail transit (LRT) was positioned as a part of the core public transportation network that supports it. The plan was revised in 2018, ten years later, but the basic city planning policy continues. Construction work on the LRT began in 2018 and will be completed in 2023, making it the first LRT to be newly developed on all routes in Japan. Eight out of the ten bases that Utsunomiya City aims to establish as a networked compact city are connected by public transportation with dedicated tracks (see Figure 6.4). A city that was overly dependent on automobiles is being reborn as a city of the future, with highly convenient public transportation at the center of the transportation system.

6.3 HEALTH-BASED CITY PLANNING

6.3.1 City and health

The maintenance of public health is the most basic element of city planning. The establishment of a water supply system for safe

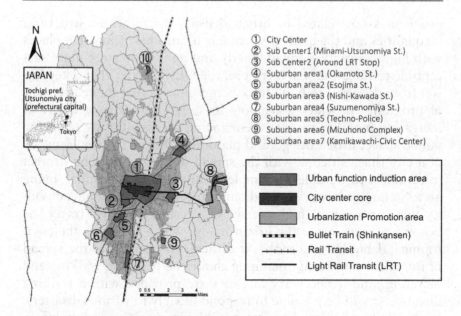

Figure 6.4 A network-type compact city in Utsunomiya.

water supply and a system to properly dispose of the large amount of garbage and waste generated by the city are the conditions for the growth of a city. In Edo (Tokyo), the sewage network and water supply system were developed from the 17th century, and the total length of the water supply system was one of the largest at the time, reaching 150 km, supporting the life of the world's largest city with a total population of over 1 million by the early 18th century. In particular, the sewage network only discharged water for domestic use and rainwater, and feces and urine were sold as fertilizer to the surrounding farmers, thus maintaining the water quality of the rivers and creating an eco-system of resource recycling. In the mid-19th century, in order to improve the cholera epidemic caused by poor sanitation, Haussmann (1809–1891) remodeled Paris, widening narrow streets to let in light and wind, and improving water and sewage systems to create a sanitary modern city.

The 2020 COVID-19 pandemic raises questions about the proper state of public health in modern cities. As a measure to prevent infection with COVID-19, which is transmitted by droplet, it was recommended to stop the flow of people and maintain a social distance. On the other hand, the vulnerability of densely populated cities has been discussed, but no direct relationship between urban density and high infection rates has been found. In "Cities Policy Responses to COVID-19", OECD (2020) notes that the health problem is not related to urban density but rather to structural inequalities and the quality of urbanization. This means that places with high concentrations of elderly and poor people are more susceptible to infection. Furthermore, the OECD states that a key lesson from the crisis to build back better cities is that the rediscovery of proximity provides a window to shift faster from a target of increasing mobility to one of enhancing accessibility by revisiting public space, urban design and planning. One of the key elements of a city that can cope with the spread of a virus through droplet infection is to be able to complete daily life in a familiar living area. So far, we have expanded our living area by developing mobility that can move at high speed. As a result, extensive travel has become a prerequisite for urban living, which increases the environmental burden and makes it difficult to respond to the spread of infection. Future city planning should be based on walking and bicycling, and if necessary facilities are provided within walking distance, it will be possible to respond flexibly to various disasters. It can also be expected to foster local communities and maintain health through increased physical activity.

The World Health Organization (WHO) defines health as "a state of complete physical, mental and social well-being and not merely the absence of disease or infirmity" and presents a broad concept. According to this definition, for people to lead healthy lives, a wide range of perspectives are required, from personal health management to improving the environment of local and urban areas.

In Japan, the following five initiatives are presented as necessary efforts for "community development of health, medical care, and welfare".

1. Raise residents' health awareness and acquire exercise habits.
2. Increase participation in community activities and revitalize activities that support the community.
3. Intentionally secure urban functions in daily life areas and walking areas.
4. Forming a walking space that encourages walking around the city.
5. Encouraging the use of public transportation.

The first item is personal healthcare, the second item addresses community issues, and the next three are all related to urban walking and traffic conditions.

Physical health is relatively easy to grasp, and traffic measures such as barrier-free measures are implemented to eliminate physical obstacles. On the other hand, it is difficult to grasp mental health, and the response to city planning is limited. However, degrading mental health can lead to suicide in the worst-case scenario. Looking at the world, 800,000 people die from suicide every year, which is a big social problem. The annual number of suicides in Japan is about 20,000 (2019), which is about six times the number of fatalities in traffic accidents making it a profoundly serious problem that must be considered. There are various causes of suicide, but as a result of investigating the relationship between the suicide rate and the urban environment in Japan, areas with less suicides are characterized by a high population density, diversity in the community, as well as more relaxed and flexible relationship between people. However, research in this field is still insufficient, and further development is expected.

6.3.2 Mobility in healthcare

Walking is a basic behavior of human daily life, and when the amount of walking decreases, it is more likely to cause lifestyle-related diseases such as obesity. It is said that the number of steps, walking speed, and gait are also related to health promotion, and movement for walking is the primary demand. The amount of physical activity that has been shown to be effective in preventing the occurrence of lifestyle-related diseases is thought to be equivalent to approximately 8,000–10,000 steps per day. In addition, as physical activity (lifestyle activity and exercise) for health promotion, "60 minutes of physical activity with an intensity of 3 METs or more every day" is suggested as a standard for those aged 18–64, and "40 minutes of physical activity every day regardless of exercise intensity" for those aged 65 and above. Metabolic equivalent (MET) is a measure of the intensity of physical activity, and three METs are equivalent to the same level of physical activity as normal walking. If a working-age person's daily life includes a 30-minute walk to work or school each way, he or she will naturally reach the required amount of exercise per day.

The traffic handled in city planning and transportation planning is mainly targeted at the movement of public spaces such as roads. Therefore, a trip, which is the unit of movement of a person, is counted when leaving the building or site, and is not counted as a trip when a person moves within the site. On the other hand, walking, which deals with health science and rehabilitation, is mainly targeted at private spaces on the premises, but is connected to public spaces.

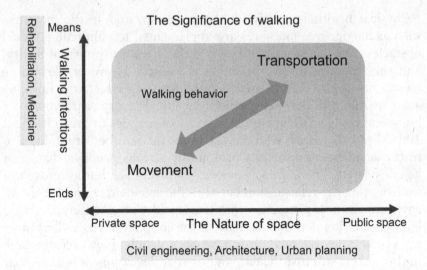

Figure 6.5 Movement and transportation in medicine.

It is also the difference between disciplines that focus on space or people. In considering the fusion of the two, it is necessary to comprehensively consider public space and private space as human activity areas. For example, rehabilitation performed in a private space seamlessly expands into a public space. The main purpose of public space is to efficiently handle transportation, which is a derived demand, but if it approaches the primary demand that makes walking fun, it will greatly contribute to healthy town development.

6.3.3 Walking distance and utility

The following is a model showing how utility changes because of walking, regardless of whether the movement is inside or outside the site (see Figure 6.6). If the distance traveled is zero, the utility is also zero. At the beginning of movement, there is a strong physiological desire to move the body, and the utility increases sharply with the movement. When a certain distance is reached, satisfaction peaks and the utility begins to decline little by little. If the movement continues, fatigue and tiredness will occur, and it will be unpleasant to move. When this happens, the negative utility gradually increases. When arriving at the destination, the integral value of the utility from the starting point becomes the utility of the entire movement, which can be said to be the satisfaction of the person. If the total value is positive, the movement can be interpreted as the primary demand, and if the total value is negative, it is the derived demand.

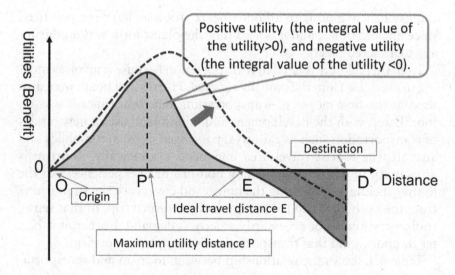

Figure 6.6 The relationship between utility and time in mobility.

In this figure, the point E where the utility of walking becomes zero is the ideal travel distance because the integrated value of the utility is maximized. If the distance to all destinations in a city is comparable to the ideal travel distance, then the city has no travel defined as derived demand. It can be said that the quality of life (QoL) of people in transportation is the largest city. However, the preference of movement is diverse, and the ideal distance traveled varies from person to person, so not all people get the maximum QoL. If the ideal travel distance is extended, there is no doubt that the number of people who feel happy in traveling will increase accordingly.

In other words, by improving the quality of the walking space and creating a fun city, people's satisfaction will increase, and it will contribute to the building of a healthy body. "Walkable" is one of the important keywords of city planning.

6.4 TOURISM AND COMMUNITY DESIGN

6.4.1 Tourism and transportation

Tourism means traveling and staying with enjoyment outside their usual environment. The word tourism is said to have first appeared in 1811, and its etymology is the Latin word tour, meaning "to turn on a lathe". The etymology of "tourism" in Japanese comes from the Chinese classic "I Ching", which comes from looking at the excellent things of the region with all one's heart. Tourism has been practiced since ancient times, and it is said that the oldest form was a pilgrimage to a sacred place.

Tourism is closely related to transportation because it involves moving to areas far from the daily living area. Horses and boats were also used as the first means of transportation, but most times it was on foot. Later, with the development and widespread use of new modes of transportation such as railways, passenger ships, automobiles and aircraft, the tourist travel area expanded dramatically. Tourism is an activity in which people enjoy non-ordinary experiences, so the journey begins with leaving the house and covers a series of journeys from the outbound trip to the stay and the return trip. In that sense, transportation is not necessarily a derived demand. Rather, it is better to understand that transportation is also part of tourism.

Table 6.1 shows the relationship between tourism and transportation by means of transportation, divided into primary demand and derived demand. Considering transportation as a means of transportation to tourist destinations, it is important to be able to travel as quickly, cheaply, and as comfortable as possible. Therefore, it is

Table 6.1 Relationship between tourism and transportation

Traffic	Derived demand	Primary demand
Meaning of the move	Means of transportation to the destination (The ends and the means are separate)	The move itself is a tourist resource (The ends and means are identical)
Railway	Fast, Comfortable, Reasonable price, Improvement of access facilities (Access to the sights)	More enjoyable way of the trip, High value-added vehicles (Integrated with tourist attractions)
Car	High standardization of roads, Improvement of vehicle performance (Smooth driving conditions.)	Scenic road maintenance, Vehicles with lodging facilities (Comfortable driving conditions)
Ships and aircraft	Larger and faster, Development of ports and airports (Coordinate with other transportation systems)	Sightseeing boats and flights, Improved livability (high value-added) (Improved visibility)

necessary to develop a transportation route that can move smoothly in large quantities. On the other hand, if the movement itself is considered a tourism resource, devising ways to enjoy it while traveling is essential. That is to add value to the vehicle itself and create an environment where tourists can comfortably view scenic places.

The number of tourists is increasing rapidly all over the world due to the development of transportation, the rising income level of people and the growing taste of tourism. According to United Nations World Tourism Organization (UNWTO) estimates, the number of overseas travelers alone amounts to 1.4 billion annually in 2018. In addition to the travel industry, the accommodation industry, and the restaurant industry, the tourism industry is a comprehensive industry with a wide range of influence, including transportation industries such as airlines, manufacturing industries such as specialty products, and entertainment and leisure in tourist destinations. Therefore, the development of the tourism industry is one of the core industries of a country that has a great impact not only on the local economy but also on national interest.

Due to the nature of the tourism industry, which is based on inter-regional exchanges, the impact of the 2020 COVID-19 pandemic is the most serious of all industries and has a major impact on other industries in the region and the country.

6.4.2 Sustainable tourism and community development

In response to the harmful effects of the growth of the tourism industry, such as environmental destruction, the issue of sustainable tourism has been on the agenda since the end of the 20th century. The tourism industry should originally consist of local resources such as the rare natural environment, but sometimes priority is given to expanding economic profits, and nature and local culture are destroyed in the process of tourism. In response to the popularized mass tourism that can threaten the sustainability of the tourist destination itself, another form of alternative tourism was proposed. For example, eco-tourism and community-based tourism have been attracting attention in recent years. For tourism to be sustainable, it is important that the economy, the environment, and society are kept in good balance.

Similar changes have been seen in Japan's tourism industry since the 1990s. The collapse of the bubble economy in the early 1990s brought to light problems such as hollowing out of the city and depopulation of mountainous areas, and the tourism industry expanded by the bubble economy was forced to shrink or withdraw. Against this backdrop, in the 1990s, "tourism-based city planning" began to flourish to integrate the previously separate activities of local community development and tourism attracting people from outside the city. Efforts to develop tourism on an internal basis based on the region's indigenous nature, rather than the exotic and exogenous development of tourism by large external capital, are highly regarded.

Yufuin, which is famous for being a successful example of tourism-based town planning in Japan, was one of the first places to implement this kind of initiative. Since the 1970s, Yufuin, once a small, desolate Japanese hot spring resort, has been engaged in tourism-based town planning with three goals: to nurture industry and enrich the town, create a beautiful cityscape and the environment, and create harmonious human relationships. It is interesting to note that this goal is the same as the three elements in sustainability (economic, environmental, and social).

In recent years, tourism-based city planning efforts have evolved in a variety of ways. For example, for tourism town development that utilizes railway resources, there has been a shift from traditional point-based activities to encouraging wider-area cooperation by making the railroad lines "one town" through PR activities to attract visitors. Alternatively, a cooperative system has been created to support regional railroads that continue to be in the red, in which the rich natural scenery can be seen from the train windows, attracting tourists. This round activity in the tourist area has been designed to maximize the primary demand of transportation. Transportation operators have also come up with a variety of plans ranging from tatami mat trains to luxury cruise trains.

The important thing is that the primary demand of travel and the primary demand of stay are combined to create an attraction for the region. Maximizing the attractiveness of both transportation and land use, fostering the human resources to be actively involved in these areas, finding the right people to work with, securing the viability of the project, and guiding the city toward sustainability. In the future, less time will be required for obligatory travel, such as commuting to work and school, and for life maintenance travel, such as shopping and hospital visits. On the other hand, the more discretionary uses of time such as recreation and entertainment will increase as well as the demand for tourism-based city planning.

6.4.3 Methods of tourism-based city planning

The content and the way of proceeding with tourism-based city planning differ greatly depending on the characteristics and current situation of the region. In developed areas where tourism has already become a mature part of the urban economy, it is necessary to take a unique approach. Based on a Japanese case study, this section presents important points to consider when starting to develop a tourist town. The goal of "tourism-based city planning" is to improve both the QoL and the revitalization of the region, and five points of view are proposed for this purpose.

Viewpoint 1: viewpoint from the outside/Viewpoint of the entire city (utilization of local resources and existing assets)
People who grew up in that city know the city well but may be biased in their views, and some aspects arise that are too obvious to notice. To attract people from the outside, it is necessary to correctly understand the interests of the outsiders and actively incorporate the perspective of the outsiders. In addition, the excavation of local resources and the utilization of existing assets require a careful review of local resources and a view of the city as a whole.

Viewpoint 2: leader (emergent human resources continue to make changes inside)
A bearer is necessary for town development. To achieve this, it is important to start with the projects that can be undertaken, and then find and develop human resources from there. Emergent human resources who carry out creative community development activities and actively disseminate information are indispensable for tourism community development.

Viewpoint 3: vision (to attract people inside and outside, to be what you want to be)
In order to create new values that will attract people inside and outside the region, it is important to set out a vision of what they want to be (a vision of the future). By setting goals that each stakeholder is responsible for realizing, and working together to achieve them, they can reach the goals of improving the QoL and revitalizing the community.

Viewpoint 4: private funding and know-how (to secure business feasibility and aim for sustainable activities)

The viability without excessive reliance on public funds is required for tourism-based city planning to be sustainable. To achieve this, it is vital that they aim for sustainable activities, utilizing private sector funding and know-how, while also aiming for projects that are appropriate to their stature.

Viewpoint 5: colleagues (finding and collaborating with others who share a common challenge)

The project will develop into a city-wide initiative, by connecting various stakeholders, including residents, businesses, and non-profit organizations. By finding friends who share common challenges and working together, urban development is sustained and leads to inter-regional cooperation.

In general, if urban development relies heavily on tourism, the rise and fall of a city will be greatly influenced by tourism trends. The approach presented here is just a guidepost for the region to coexist with tourism in an independent manner. It is important to maintain a balance between the two to attract people from the outside while promoting the local economic production for local consumption.

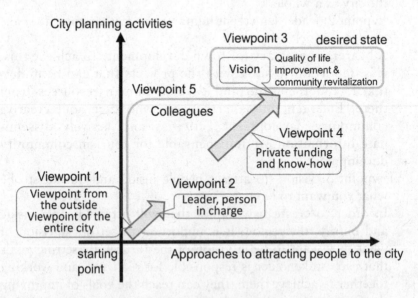

Figure 6.7 Points of view for tourism-based city planning.

Source: Modified from City Bureau, Ministry of Land, Infrastructure, Transport and Tourism. *Guidelines for Tourism-based City Planning* (2016).

REFERENCES

City Bureau, Ministry of Land, Infrastructure, Transport and Tourism.2014. *Guidelines for the Promotion of City Planning for Health, Medical and Welfare* (in Japanese). https://www.mlit.go.jp/toshi/toshi_machi_tk_000055.html (accessed August 15, 2020).

City Bureau, Ministry of Land, Infrastructure, Transport and Tourism. 2016b. *Guidelines for Tourism-based City Planning.* https://www.mlit.go.jp/toshi/kanko-machi/index.html (accessed August 15, 2020).

Masayuki Doi, Noboru Sakashita. 2002. *Transportation Economics,* Tokyo: Toyo Keizai Inc. (in Japanese).

Noboru Harata (ed.). 2015. *Transport-Based City Planning: Challenges from Local Cities,* Tokyo: Kajima Institute Publishing Co., Ltd. (in Japanese).

Japan Society of Traffic Engineers. 2006. *Transport-Based City Planning: Learning from Cities in the World and Japan,* Tokyo: Maruzen Publishing Co., Ltd. (in Japanese).

Shigeru Katsuta (ed.). 2015. *Introductory Exercise Physiology,* 4th ed., Tokyo: Kyorin Shoin Co., Ltd. (in Japanese).

Noboru Kawazoe. 1982. *The City from the Other Side: A Life History,* Tokyo: Japan Broadcast Publishing Co., Ltd. (in Japanese).

Ministry of Health, Labour and Welfare in Japan. 2013. *Physical Activity Standards for Health Promotion 2013.* https://www.mhlw.go.jp/stf/seisakunitsuite/bunya/kenkou_iryou/kenkou/undou/index.html (accessed August 15, 2020).

A. H. Maslow 1943. A Theory of Human Motivation. *Psychological Review,* 50, 370–396.

A. H. Maslow 1954. *Motivation and personality.* New York, NY: Harper & Row. doi:10.1037/h0054346.

Patricia L. Mokhtarian, Ilan Salomon, Matan E. Singer. 2015. What Moves Us? An Interdisciplinary Exploration of Reasons for Traveling, *Transport Reviews,* 35(3):250–274.

Masahiro Nei. 2018. *A History of Economics in the Original English Language,* Tokyo: Hakusuisha Publishing Co., Ltd. (in Japanese).

Yukio Nishimura (ed.). 2009. *Tourism Town Development: Regional Management Starting from Town Pride,* Kyoto: Gakugei Shuppansha (in Japanese).

OECD. 2020. OECD Policy Responses to Coronavirus (COVID-19). *Cities Policy Responses.* http://www.oecd.org/coronavirus/policy-responses/cities-policy-responses-fd1053ff/ (accessed August 15, 2020).

Katsutoshi Ohta. 2008. The Evolution, Challenges to and Direction of "Transport-based Machizukuri (Community Development)", *IATSS Review,* 33:136–139.

Mayumi Oka. 2013. *A Pleasant Town to Live in: There Is a Reason for This Low Suicide Rate,* Tokyo: Kodansha Ltd. (in Japanese).

Takeshi Osawa, Seiji Yoneda. 2019. *Yufuin Model: Tourism Strategy Through Innovation That Takes Advantage of Regional Characteristics*, Kyoto: Gakugei Shuppansha (in Japanese).

S. Reichman. 1976. Travel Adjustments and Life Styles – A Behavioral Approach In *Behavioral Travel-Demand Models*, eds. P.R Stopher, A.H Meyburg, 143–152. Lexington, MA: D.C. Heath and Company.

Hiroshi Suda. 2003. *New Tourism Resources Theory from a Practical Point of View*, Tokyo: Transportation News Co., Ltd. (in Japanese).

United Nations World Tourism Organization (UNWTO). 2018. *World Tourism Statistics*. https://www.unwto.org/global/press-release/2019-01-21/international-tourist-arrivals-reach-14-billion-two-years-ahead-forecasts (accessed August 15, 2020).

Yumi Vincent Fujii, Kiyohito Utsunomiya. 2016. *Why There Are No Shuttered Streets in Provincial Cities in France*, Kyoto: Gakugei Shuppansha (in Japanese).

World Health Organization (WHO). 2020. *Mental Health and Substance Use, Suicide Data*. https://www.who.int/mental_health/prevention/suicide/suicideprevent/en/ (accessed August 15, 2020).

Hirotaka Yamauchi, Kenzo Takeuchi. 2002. *Transport Economics*, Tokyo: Yuhikaku ARMA (in Japanese).

Chapter 7

Advanced transport

7.1 THE BIRTH OF ADVANCED TRANSPORT

7.1.1 Thinking about new transportation

New transportation exists in every generation. In the 19th century, the railroad was the new transportation, and in the 20th century, the automobile became widespread as the new transportation. So then, what is the new transportation of the 21st century today? Assuming that the new transportation was born out of people's life and demand from the past, part of the clue lies in history. Going back in time about half a century, and focusing on the 1960s, when automobile traffic was at its heyday, we will explore under what circumstances and what kind of future traffic people were anticipating at that time.

In his book "New Movement in Cities", B. Richards (1966) discussed the transportation of the future, mentioning the transportation proposed over the past 100 years. Since future urban forms will be influenced by new technological developments in the field of transportation, desirable cities will differ according to basic hypotheses about transportation. In particular, he mentioned two different possibilities with the use of private cars as the main means of transportation. One was the spread of automobile traffic, assuming that 75% of all families in Europe and the United States would have a private car by the year 2000. It was possible to plan a corresponding, fully motorized city, but this would require countries with ample economic resources and very large areas of land for streets and parking. The other stated that in a densely populated country, the most legitimate solution is to consider the balance between private and public transportation, and that the relationship between land use and transportation needs to be considered. This concept is still applicable today. Moreover, Attractive public transportation is important for coexistence with automobiles, and he presented a park-and-ride system for coordination between automobiles and public transportation. Comparing the two, the former is inefficient in its land use, and as a realistic future city, it needs a secondary transportation system to supplement the mainline public transportation system. Secondary transportation systems were non-stop moving transportation systems, pedestrian conveyors that serve as transportation nodes and elevation transfers, mini-rail systems used in exterior spaces, and mini-cars for private use. Although some of the proposed transportation systems have not yet been put to practical use, the hierarchical and interconnected nature of transportation systems is still the basis for the construction of transportation systems.

In the prologue of his book "Transportation in the World of the Future", H. Hellman (1968) described the ideal of a fully integrated public transport system in a century's time that would meet people's mobility needs. He lamented that in the United States at the time, various transportations had developed, but they competed with each other, lacked coordination, and were not integrated into what could be called a transportation system. He also pointed out that the average American household spends 13% of its annual income on transportation, that half of the central business district (CBD) is taken up by streets, sidewalks and parking, and that 50,000 people are killed in car accidents every year. Transportation systems emerged, but they were expensive and difficult for users to use because they did not cooperate with each other to achieve profitability on their own. As a solution to this problem, he hoped for the construction of highways, but noted that automobiles alone would not solve the problem, and he envisioned a form in which railways and automobiles were best combined through technology. One of them is to automate the highway. The automatic control of vehicles by computer and electronic means would increase traffic capacity and improve vehicle comfort and safety. The vehicle-controlled road near the center line was called the "Autoline". It also mentions the importance of high-speed rail between cities and subways and monorails within cities. At the time, he referred to the success of the latest Japanese bullet train (1964) as an important issue as to whether it could succeed in the United States, which is increasingly dependent on automobiles. He also mentioned the development of magnetic levitation trains and linear motor technology for even higher speed trains. For intra-city transportation, he described the applicability of electric vehicles, automated minibuses, and moving sidewalks, and introduced the "people mover", a fixed-route transportation system using small vehicles. In order to link roads and railroads, he stated, the idea is to combine collector systems (cars, buses, subways, etc.) with mainline railways. It is very interesting to find that the new transportation systems predicted here include many that are already in operation or under development.

What can be gleaned from these books is that they almost correctly foresaw the current urban transportation problems and had already proposed transportation technologies to solve them half a century ago. Unfortunately, the urban transportation problem has not yet been solved, but the direction of the solution is almost the same. Just as the transportation that was considered ideal half a century ago is gradually becoming a reality today, it may take half a century to realize the ideal image that is currently being considered. The advanced transport and city planning discussed in this chapter also consider the realization of a better transportation society by following the wisdom and experience of our predecessors and making full use of the transportation technologies that have become reality.

7.1.2 What is the advanced public transport?

Advanced transport is a generic term used to describe transportation based on new technologies, and continues to evolve with the various technological developments, both in public transport used by an unspecified number of people, and in private transport used by specific individuals. It will begin with an overview of the new transit system, which has received a lot of attention in the second half of the 20th century as the advanced public transport. The term "new transit system" is narrowly defined to refer to the Automated Guideway Transit (AGT), which is an automated, dedicated track system, but in a broader sense, the term is used to describe all public transportation systems developed with new technology. Typical examples include monorail, linear motor cars, light rail transit (LRT), bus rapid transit (BRT), and demand responsive transport (DRT), in addition to AGT.

AGT or a similar system, APM (Automated People Mover), was created in the 1970s to provide transportation between mass transit (e.g., railways) and medium-volume transit (e.g., trams and buses), and to ensure regularity with dedicated tracks. It is a new type of public transportation that is small, lightweight, and automatically driven by a computer, and has been promoted in many countries around the world. They were especially popular for access to airports, large new towns, and event venues. The intersections with existing roads require overhead or underground level crossings, which made it relatively easy to introduce them to new areas, but not so easy to introduce them to existing areas.

The introduction of new public transport in the existing urban areas, especially in the city center, will begin to be considered with the re-division of road space. Against the backdrop of the decline of urban centers due to excessive dependence on the automobile, medium-volume intra-urban public transportation was sought as a means of revitalizing the city's urban centers from around the 1970s in the United States. The focus was on systems such as LRT and BRT, which evolved from conventional trams and buses. LRT is an abbreviation for the Light Rail Transit, which uses low-cost, light rail cars instead of heavy rail cars like conventional railways. The LRT first opened in 1978 in Edmonton, Alberta, Canada, and has spread around the world in conjunction with the city center revitalization program. The design and construction of the Edmonton line was characterized by an emphasis on simplicity, which helped to keep construction costs down and provide facilities that could be operated efficiently and reliably, features that were carried over to the new light rail system that followed. In addition, BRT, which enhances punctuality and transit capacity with dedicated tracks and connected vehicles, has begun to be introduced around the world. The first city to introduce BRT is said to be Curitiba (Brazil), which opened in 1974. What LRT and BRT have in common is that they both utilize the characteristics of existing trams and buses, while ensuring speed and punctuality by building dedicated routes in congested urban areas. Compared to LRT, BRT is relatively cheaper to install and is widely used as an intra-city

Light Rail Transit

Strasbourg (France)

Bus Rapid Transit

Curitiba (Brazil)

Figure 7.1 LRT and BRT.

public transportation system in developing countries and regions with relatively low demand. Another attraction of BRT is its ability to run on ordinary roads in addition to dedicated roads.

7.1.3 What is the advanced private transport?

Since the 1990s, Intelligent Transport Systems (ITS) have emerged as a mobility system that makes advanced use of information technology (IT). ITS refers to a new transportation system that aims to solve road traffic problems by networking people, roads, and vehicles with information using the advanced information and communication technology. In addition to the development of conventional vehicle-centric technologies, its distinctive feature is its ability to receive and transmit information to and from people and roads, which has been put to practical use in many countries.

With regard to technological innovation in vehicles, the development and practical application of more efficient vehicles, such as hybrid and electric vehicles, which are powered by both gasoline and storage batteries, began at the end of the 20th century in response to growing environmental concerns. At the beginning of the 21st century, the performance of batteries improved significantly, contributing to the spread of electric vehicles and fuel cell vehicles (FCVs) that generate electricity using fuel such as hydrogen.

While motorcycles and electric (assisted) bicycles have been the most common means of transportation for individuals, new mobility-assistive devices have been gaining attention in the 21st century. In particular, personal mobility is a new type of small, single-seat mobility device, in which companies have begun to develop and offer a wide variety of models to the market. The first of these was the Segway, announced in 2001. Due to the ease of operation and the flexibility of the vehicle's operation for anyone to use, it was used especially for event venues and for patrolling and security in the city. On the other hand, the response to the driving route differs depending on the road traffic system of each country. In Japan, for example, the use of the vehicle was either modified to comply with the Road Traffic Law or limited to use on non-roadways.

The micro mobility is smaller, lighter, and slower than personal mobility and is mainly used as a short-range mobility aid, and its use is expanding. Traditionally popular shared bicycles, electric bicycles, or electric kick scooters are beginning to be introduced as a shared system to support short distance travel, especially in urban centers around the world.

With such a wide variety of transportation systems being developed and put to practical use, self-driving technology is one of the most anticipated new technologies. Compared to conventional automobiles, which have rapidly spread around the world in the 20th century due to their usefulness and comfort, this new technology further enhances the superiority of automobiles by reducing traffic accidents caused by human error and effectively utilizing travel time by freeing people from the driving tasks. There are six levels of automation for autonomous cars, ranging from Level 0 to Level 5, according to the 2016 SAE International definition (SAE J3016).

Level 0 (No Automation): the human driver controls all driving tasks.

Level 1 (Driver Assistance): the system controls a specific function (such as steering or accelerating).

Level 2 (Partial Automation): the system controls both steering and acceleration/deceleration.

Level 3 (Conditional Automation): the system can drive automatically but may require the human driver's attention and response.

Level 4 (High-level Automation): the system is fully autonomous within the operational design domain.

Level 5 (Full Automation): the system is fully autonomous in all situations.

As of 2020, vehicles up to the Level 3 will be available for sale. But the widespread adoption of self-driving vehicles at Level 4 and above will require further progress in self-driving technology, the development of related legislation, a review of the insurance system, and a review of existing roads and other infrastructure. In addition, the areas expected to be at the center of technology development in the automotive industry in the future are Connected, Autonomous, Shared & Services, and Electric, which are known as CASE.

With the diversification of land transportation within cities, the use of short-distance delivery and movement using the upper space of cities is also being considered with drones beginning to be used as unmanned aerial vehicles (UAVs). However, there are many issues to be addressed, such as the large amount of energy required for transportation, safety, and securing locations for takeoff and landing for practical use.

7.2 THE ROLE OF ADVANCED TRANSPORT

7.2.1 Hierarchy of urban transportation and advanced transport

A variety of advanced transport systems are emerging, and future transportation systems will require the construction of comprehensive mobility system that includes mutual cooperation. For this reason, it is important to consider how to share the role with the existing urban transportation networks. In general, the hierarchy of urban transportation can be shown in Figure 7.2. The left side of this hierarchy pyramid structure shows the road network, and the right side shows the public transport network. Travel speed decreases from the top to the bottom of the pyramid structure, with inter-city, intra-city and intra-district transportation playing a role, respectively.

Primary transport (inter-city transport): expressways, major arterial roads/high-speed trains, and railways

Secondary transport (intra-city transport): arterial roads, subways, LRT/BRT, trunk buses

Tertiary transport (transport within the district): local roads/community buses, DRT, and taxis

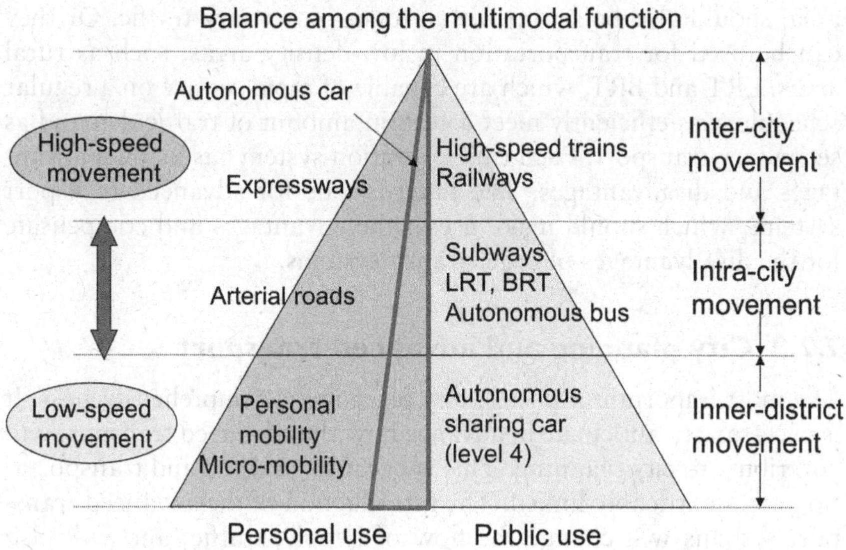

Balance among the multimodal function

Figure 7.2 Hierarchy of urban transportation and advanced transport.

The public transportation network on the right side of this figure is devastated in many overly automobile-dependent cities. Not only this, the spread of automobiles with inadequate development of highways and arterial roads will result in the overflow of cars on the roads, causing severe road congestion and traffic accidents. Therefore, it is important to construct a hierarchy of transportation in the city for the entire transportation system. If this is not possible and only localized improvements are made to congested roads and intersections, the congested areas will only be relocated, which will not lead to radical improvements. Widening narrow residential roads and constructing sidewalks has the risk of attracting passing traffic.

Therefore, it is desirable to draw an ideal transportation system for the future and to introduce advanced transport to complement and guide it. In other words, the elements required for the advanced transport systems can be summarized as follows: (1) comfort (barrier-free, low vibration), (2) efficiency (environmental performance), (3) safety (use of ICT), (4) consistency (the role of transportation systems and design), and (5) comprehensiveness (coordination with city planning).

The role of advanced transport in the transportation hierarchy should be examined to maintain consistency. For example, tertiary transport (intra-district transport) is suitable for personal mobility and micro-mobility, which are mainly for low-speed, short-distance travel. Level 4 of autonomous cars (automated driving in a limited area) should also be set up with a focus on tertiary traffic. Or they can be used for transportation in low-density areas, such as rural areas. LRT and BRT, which are capable of rapid service on a regular schedule, can efficiently meet a certain amount of traffic demand as secondary transport. Each transportation system has its own advantages and disadvantages. The same is true for advanced transport systems, which should make use of the advantages and compensate for the disadvantages of other transit systems.

7.2.2 City planning and advanced transport

The most important aspect of city planning is comprehensiveness. It is necessary to anticipate in advance how the advanced transport will contribute to city planning. This is because land use and transportation are inextricably linked. The introduction of the advanced transport systems will change the flow of people, traffic, and will also affect land use. Therefore, the future urban structure of the city can be guided by skillfully incorporating advanced transport systems.

For example, the advanced transport can contribute to the expansion of urban areas when the population is growing, and it can also play a role in guiding land use toward intensive centers when the population is declining. Therefore, the introduction of advanced transport systems requires a series of processes, including planning, project implementation, and construction, which should be carried out after careful consideration of their compatibility with the city planning master plan that shows the future vision. In particular, the advanced transport has special characteristics that are not present in existing transportation systems, so it is necessary to carefully understand the impact it will have on its surroundings.

The role that the advanced transport should play in a city depends on the characteristics of the city and its social context. Therefore, it is not possible to specify the relationship between cities and the advanced transport in a uniform manner, but as an example, one of the ideal relationships between the two will be presented for the case of constructing a compact city.

An ideal situation regarding primary transportation for inter-city travel is to have railway stations in the city center, and highway interchanges in the suburbs. The public transport network would gradually reduce the speed of travel from the city center to the suburbs, from secondary to tertiary traffic, and increase route flexibility. On the other hand, the road network will be arranged as secondary and tertiary traffic from the suburbs to the city center to reduce travel speeds, and inflow restrictions will be established in the city center. This will create an urban center based on walking and public transportation. Public transport hubs with large transport capacity (primary transport) can support city center development and transit-oriented development (TOD) because of their strong land use inducement effect. Meanwhile, the area around the highway interchange is highly suitable for truck transportation, and so it will be developed as a distribution hub such as distribution operations center, as well as an industrial base for factories and other industries. Besides, land use regulations are needed around suburban arterial roads to discourage uncontrolled development.

Secondary transportation plays a role in forming the urban axis, which represents the framework of the urban structure. It is a highly transportable network that connects the city center with various locations and is the center of intra-city mobility. The next generation of public transportation includes LRT, BRT, and self-driving buses. Secondary transport nodes form the hub of everyday life within the city and act as a hub and spoke to the surrounding area.

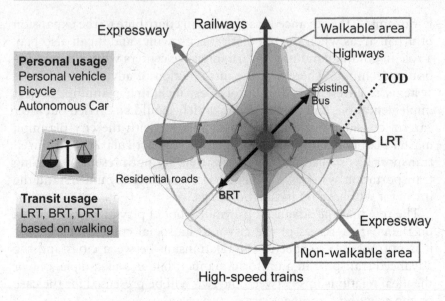

Figure 7.3 Urban structure and transportation network.

Tertiary transportation is the access transportation to each facility. While walking and bicycles are the basic means of transportation, personal mobility use depends on the size of the area. Tertiary traffic is for living spaces with speed limits of 30 km/h or less than 20 km/h and is also a pedestrian-vehicle coexistence road. They are characterized by the flexibility to change with the increase and decrease of the urban population.

7.3 CHALLENGES IN INTRODUCING AND PROMOTING ADVANCED TRANSPORT

7.3.1 Barriers to the introduction of advanced transport

In order for advanced transport to become widespread throughout society, several conditions need to be satisfied in addition to developing a new mode of transportation. Many technologically new transportation systems have been invented and some have reached the level of practical use, but in many cases, they have only been used temporarily. Therefore, even with better transportation systems than in the past, there are many challenges to the introduction and diffusion of advanced transport. In general, the challenges can be broadly summarized into three categories: institutional barriers, economic challenges, and coordination with vested interests.

1. Institutional barriers

For advanced transport to be approved, the safety of the transportation system and its affinity with the traffic community must be comprehensively assessed and legally stipulated in the Road Traffic Act and other laws. Particularly for use on public roads, a full review is carried out from a road safety perspective and, in some cases, modifications will be required to conform to existing legislation. For example, in many cases personal mobility system such as "Segway" is not allowed on public roads in the country's legal system and driving them requires modifications to meet existing safety and security standards. Autonomous vehicles (Level 4 and above) without drivers are not stipulated in the current system and will need to be revised to run on the public road. In addition, a series of systems need to be put in place to deal with the issue of liability when the advanced transport causes accidents and the insurance system to support them, which will require considerable time and public recognition to spread.

2. Economic challenges (profitability)

Even if the advanced transport is developed and permitted to operate, there must be sufficient benefits in relation to the costs for its diffusion. In private transportation, individuals make decisions about purchasing the advanced transport by comparing the benefits of using that transportation to the total cost of purchasing a vehicle, fuel, and storage costs. In public transportation, transportation operators and management entities will make similar decisions to implement the advanced public transport when the benefits are expected to outweigh the costs.

On the other hand, the spread of advanced transport systems will often require not only vehicles, but also infrastructure facilities to support them. For example, the development of charging stations and other facilities is essential for the spread of electric vehicles. The LRT will also require a large amount of money to develop dedicated tracks and stations. Therefore, the key to widespread adoption is the ability to afford the initial cost of introduction. When the advanced transport generates social benefits such as congestion reduction, environmental impact reduction and accident reduction, it becomes justified for the government and other public entities to spend some of the initial and operational costs. In this case, the public interest of the project must be determined through a certain consensus building process.

3. Coordination with vested interests (consensus building)

The genesis of the advanced transport is its superiority over existing transportation systems. The development of the advanced transport system starts when it has some advantageous characteristics over the existing transportation system. However, even when the advanced transport is developed, it faces a battle for getting users against competing existing transit systems to gain widespread adoption. Even if a new system has superior performance on its own, it is a significant hurdle to compete with existing systems that have provided transit service in the area for many years.

Individual purchasing trends are key to the spread of the new private transport system. The difference in function between advanced transport and existing transport is expressed as a price, and consumers judge the reasonableness of that price difference. If the benefits of advanced transport exceed the quoted price, more people will buy and, as a result, advanced transport will become more popular. However, this can be a matter of life and death for industry groups that make their living from existing traffic, and so there can be a variety of resistance, including political pressure.

While the basic mechanism for public transport is the same, the coordination way has a slightly different aspect. It is difficult to compete fairly, especially when there is a mix of private and public institutions in the market. In addition, because public transportation itself is a means of transportation with high public interest, even private operators are often subject to governmental permits or reports for a wide range of things, including route entry and withdrawal and frequency of operation. This market intervention by public authorities has also led to deregulation, partly due to criticism that it impedes free competition. However, this has led to an overweighting of public transport services to areas of high usage and an increase in areas of public transport disadvantage. Given this history, market intervention by public authorities requires extremely difficult decisions. In many cases, subsidies including initial investment costs from public institutions are essential for the spread of advanced public transportation, which can easily create a structure of conflict between public institutions and private operators.

In order to resolve such conflicts with existing traffic, it is necessary to have a comprehensive management system to coordinate the entire transportation system, which combines next-generation and existing traffic, rather than just leaving it to market mechanisms.

7.3.2 Strategies for implementing advanced transport

In general, the three overlapping issues of institutional aspects, profitability, and coordination of consensus building have become constraints on the spread of advanced transport. If advanced transport is deemed necessary in city planning, it is vital that the government and the private sector work together to solve these issues. In addition to coordinating between transportation systems, a comprehensive strategy that combines transportation and land use should be adopted, since changes in the transportation system will affect land use. Although it is difficult to express a comprehensive strategy in a uniform way due to the differences of national systems and customs, a long-term strategy based on the Japanese system is generally presented in the following manner.

1. Articulate the planning concept in the highest level of comprehensive planning in the city.
   For government agencies, there are a wide range of fields, such as education, medical care, and welfare, in addition to the city planning field, which are affected by the budget allocation to each other. Therefore, it is necessary to clarify the purpose and necessity of city planning in the comprehensive plan that includes other fields to ensure sustainable budget allocation.
2. Prepare the city planning master plan that clearly outlines the future vision.
   The conceptual comprehensive plan will be incorporated into a more detailed urban plan that paints a future picture of the city. In some countries, temporary political and economic trends can have a significant impact on city planning, but since city planning is essentially a long-term project with a ten-year horizon, it is desirable to have a system that is not excessively linked to short-term fluctuations. Japan's city planning master plan is a system established under the 1992 amendment to the City Planning Law, in which municipalities formulate a vision for their future cities including land use and transportation every 10 years.
3. Develop land use and urban transportation plans in line with the city planning master plan.

Develop land use and urban transportation plans with the interrelationship of land use and transportation in mind. In particular, advanced transport will affect future land use, and therefore land use should be regulated and guided to the desired areas defined in the future vision. In Japan, this applies to the location normalization plan under the amendment to the Act on Special Measures Concerning Urban Renaissance (2016) and the regional public transport network formation plan under the amendment to the Act on Revitalization and Rehabilitation of Local Public Transportation Systems (2014).

4. Implement transportation projects while providing land use regulation guidance.

   Unregulated development should be discouraged, while development should be guided in the right places. Urban development should focus on private sector projects, while the government should use subsidies and other resources to guide projects appropriately. In addition, if the initial cost of a transportation project is so high that it cannot be made independently profitable, it will be necessary to separate infrastructure development and operations. In the case of major changes to existing traffic flows, such as downtown inflow restrictions and transit malls, the appropriateness and social acceptability of transportation policies should be determined by transportation social experiments.

5. Ex-post evaluation and scientific verification of policies

   The larger the scale of a policy's implementation, the more problems it creates. Changing land use and traffic flow is also about changing vested interest structures, which can make the opposition as strong as those in favor. A comprehensive ex-post evaluation is needed, with due consideration for the response and compensation of those who have suffered losses. Particularly in the case of public projects, verification using scientific data is essential for continuity, improvement and review, and the process must be transparent and accountable.

7.4 CITY PLANNING FOR ADVANCED TRANSPORT

7.4.1 Road space for advanced public transport

In order to create a city for the advanced transport system, not only the travel space but also the entire urban space, including facilities along the roads and around the railway stations, should be designed and adapted to the next generation transportation system. City planning is a decade-long plan, and to align with the introduction of advanced transport, it must be harmonized more than a decade before its introduction. Similarly, the design standards for roads must be discussed early and revised to meet the timeline for introduction or diffusion.

The first step is to reallocate road space and rethink parking to accommodate the advanced public transport. The ways to secure dedicated travel lanes in the city center are to widen existing roads, reallocate lanes, or utilize unused surface and underground space. In general, road widening takes considerable time and money to build consensus along roads, and the use of above- and below-ground space requires huge construction costs. To proceed with the project while preserving the streetscape along the roadside, the study should begin by aiming to redistribute road space. Looking at advanced examples, many cities have introduced transit malls with the advanced public transport systems such as LRT and BRT that utilize road space to revitalize urban centers. A transit mall is a pedestrian-transit coexistence street that restricts the encroachment of private vehicles and combines public transportation (transit), such as buses and trams, with pedestrian-only streets (mall). It was first launched in 1967 at the Nicollet Mall in Minneapolis, USA. Since then, more and more cities around the world have introduced transit malls.

The introduction of transit malls that discourage the influx of private vehicles and prioritize public transportation in urban centers is one of the most important policies to build a better relationship between land use and transportation. The key to success is the ability to promote consensus building for this purpose. All manner of preliminary evidence of traffic facilitation and increase in visitors will be required to be proven in advance to gain the agreement of existing road users and roadside shoppers. Traffic facilitation requires securing alternative transportation routes, and in order to increase the number of visitors, city center spaces need to be remade to be more attractive. The key to this is how to reuse the spaces that have been occupied by cars, such as driveways and parking lots.

7.4.2 Road space for autonomous cars

The spread of autonomous cars requires appropriate responses to urban facilities. Considering urban infrastructure for automated vehicles, the driving environment of the road tends to be focused on, but the access environment and parking environment are important

Open cafe that occupied the streets
leading to the transit mall

Transit Mall (Bahnhofstrasse in Zürich)
main downtown street, speed: 30 km/h

Figure 7.4 Transit mall.

from the viewpoint of urban planning. This is because while the driving environment can be solved by advances in automated driving technology, the boarding and alighting areas and parking lots need to be maintained by the city. Until now, when using a private car, one had to get on at the parking lot of the departure point and get off at the parking lot of the arrival point. However, autonomous cars at Level 4 and above can be automatically dispatched from the nearest parking lot and automatically go to the parking lot once they get off. Or, if an autonomous car is shared, the vehicle is automatically dispatched to the next ride point. In both cases, parking would not need to be provided in the vicinity of the destination and parking capacity could be reduced. Instead, adequate boarding and alighting areas would be needed more than ever before.

The need for adequate boarding and alighting areas is also due to the problems with the use of the road shoulder (curbside) between the roadway and the sidewalk currently. Although road traffic laws prohibit boarding and alighting near intersections, crosswalks, and bus stops, in reality, boarding and alighting are frequently seen on the frontages of destinations. It is understandable when you consider the current use of taxis, but inappropriate riding on the curbside is tacitly tolerated and violations of road traffic laws become an everyday affair in urban areas. Meanwhile, because autonomous vehicles comply with the law, they cannot get in and out of the car illegally, as is the case in the past. In other words, to accommodate an automated society, the design standards for the curbside between the roadway and the sidewalk need to be reviewed, including the placement of legal automated passenger spaces in the roadway. In addition, proper management of curbsides is also important.

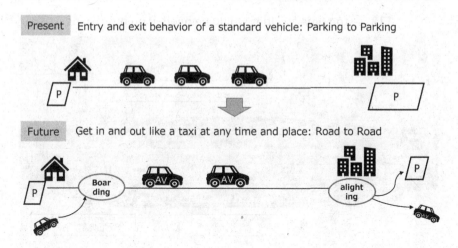

Figure 7.5 Changes in boarding and alighting locations for automated vehicles.

Also, a driveway apron should be provided on the side of the property to get in and out of the vehicle safely without affecting road traffic. Alternatively, it is ideal to build a side road as shown in Figure 7.6. Even today, porte-cocheres are permanently installed in luxury hotels, but most facilities do not have them. Side roads can be installed if planned in conjunction with a new road, but it is difficult to install them later. Therefore, the mandatory attachment standards

Figure 7.6 Example of side roads for automated driving.

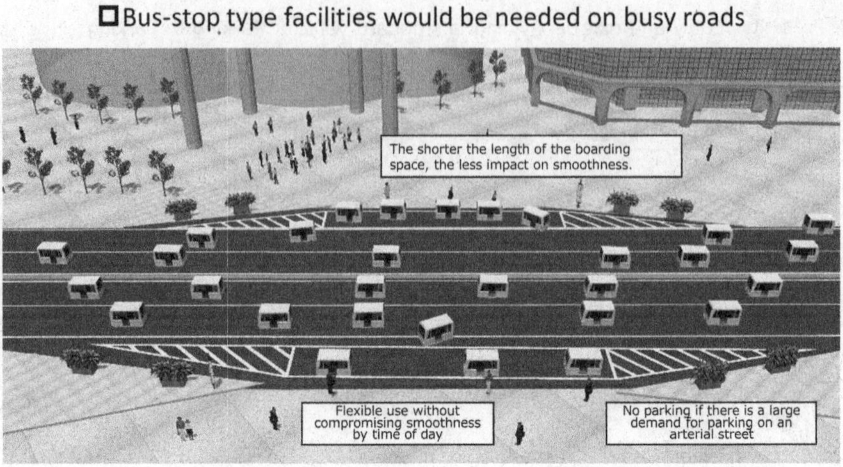

☐ Bus-stop type facilities would be needed on busy roads

The shorter the length of the boarding space, the less impact on smoothness.

Flexible use without compromising smoothness by time of day

No parking if there is a large demand for parking on an arterial street

Figure 7.7 Designing passenger spaces for arterial streets.

for roadside facilities and road design standards need to be reviewed as soon as possible to meet the needs of an automated society.

7.4.3 A scientific approach to implementation

Scientific support is needed to move forward with these plans. A preliminary understanding of the impact of the introduction of advanced transport on cities will provide a basis for making various policy decisions. This section describes a simulation of a future city that focuses on five items: economy, environment, and landscape in addition to land use and transportation. Each of these simulations has its own interrelationship with the other. The output values of the land use simulation become the input values for the traffic simulation, and the output values are used to analyze not only the congestion level but also the project profitability and estimate the environmental impact. The result of the synthesis of all these factors leads to the landscape simulation scenario.

1. Land use simulation: projection of future population distribution and urban structure
   The introduction of advanced transport changes location potential, which in turn affects land rents and changes the population distribution in the future. The changed population distribution will also affect the future demand for transportation. For the estimation of future land use, there is the land

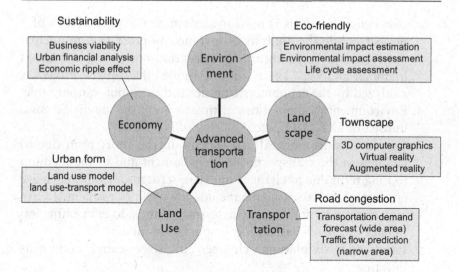

Figure 7.8 Advanced transportation and urban simulation.

market equilibrium model and the spatial interaction model. Additionally, methods for analyzing the relationship between land use and transportation include the integrated land use-transport model (ILUT) and the computable urban economic model (CUE).

2. Traffic simulation: traffic congestion forecast, traffic demand forecast

As new modes of transport are created, people's transport behavior changes. The future OD traffic volumes by mode of transport and traffic volumes per road link can be calculated by forecasting traffic demand over a wide area. In addition, more detailed traffic flow forecasts will be conducted for targeted areas to predict the extent and location of future congestion. Traffic demand forecasting in wide area traffic is mainly the classified aggregate model such as four-step estimation method and the disaggregate behavioral model. As the traffic flow prediction in narrow area traffic is calculated with micro traffic flow simulation.

3. Economic simulation: project profitability, urban finances, urban economy

Analyzing project profitability, such as construction costs and project revenues at the transportation project unit, and the impact of changes in urban structure on city finances or calculating the economic spillover effects of the city is crucial. The cost-benefit analysis is used to determine the profitability of a project, while the analysis of income (property tax, municipal tax, etc.) and expenditure (construction cost, etc.) is done as financial analysis. The economic ripple effect of the project is analyzed by the econometric model and the input–output table.

4. Environmental simulation: estimation of the environmental burden of CO_2, NO_x, etc.

   Conduct environmental assessments in the short term due to changes in the transportation environment and in the medium to long term due to changes in urban structure. Comprehensive environmental assessments include life cycle assessment (LCA), and environmental impact assessments include environmental impact estimates based on transportation models.

5. Landscape simulation: changes in streetscape, consensus building

   As the introduction of the LRT and other advanced transport systems will significantly change the cityscape, the use of CG and VR will be helpful to recreate the future urban landscape for coordination and consensus building among stakeholders. Reproducing urban space with 3D-CG models facilitates the provision of information through video, and reproduction of

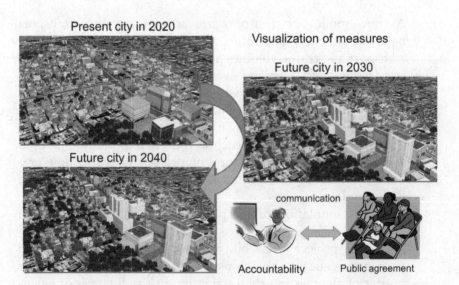

Figure 7.9 Landscape simulation of future urban space.

urban spaces through virtual reality VR or augmented reality AR enables people to experience simulated urban spaces.
Changes in the traffic environment have diverse impacts on society, such as the risk of road accidents, causing fluctuations in the number of vacant houses and vulnerable road users, and gentrification through the improvement of living spaces. The analysis from a variety of approaches is desirable to foresee what kind of society will emerge in the future.

REFERENCES

Robert T. Dunphy, Robert Cervero, Frederic C. Dock, Maureen McAvey, Douglas R. Porter, Carol J. Swenson. 2004. *Developing Around Transit: Strategies and Solutions That Work*, Washington, DC: Urban Land Institute.

Sanshiro Fukao. 2018. *Mobility 2.0,* Tokyo: Nikkei Publishing Inc. (in Japanese).

Hal Hellman. 1968. *Transportation in the World of the Future*, New York, NY: M. Evans and Company, Inc.

William D. Middleton. 2003. *Metropolitan Railways: Rapid Transit in America*, Bloomington, IN: Indiana University Press.

Akinori Morimoto. 2019a. Advanced Transportation and Compact Cities in Rural Cities, *Urban Renewal*, 594:41–44 (in Japanese).

Akinori Morimoto. 2019b. Future City Planning With Autonomous Vehicles, *The "Jutaku" A Monthly of the Housing*, 68:14–18 (in Japanese).

Brian Richards. 1966. *New Movement in Cities*, New York City: Sterling + Publishing Company.

Society of Automotive Engineers. 2018. *SAE J3016 Taxonomy and Definitions for Terms Related to Driving Automation Systems for On-Road Motor Vehicles.* https://www.sae.org/standards/content/j3016_201806/ (accessed August 15, 2020).

Koki Takayama, Shun Okano, Akinori Morimoto. 2020. Research on the Curbside Focusing on Getting on and off Environment of the Autonomous Driving Vehicle, *Journal of Japan Society of Civil Engineers*, D3-75-6:I_565–I_574 (in Japanese).

Vukan R. Vuchic. 1999. *Transportation for Livable Cities*, New Brunswick, NJ: Center for Urban Policy Research.

Chapter 8

Cities and logistics systems

8.1 LOGISTICS PLANNING IN CITIES

8.1.1 Definition of logistics and its function

One of the crucial factors for the establishment of a city is how efficiently it can continue to be supplied with a variety of goods from its surrounding area. Therefore, cities have been formed in locations convenient for shipping and riverboat transportation of large amounts of goods since ancient times. This section outlines the basics of logistics and its relationship to the city.

The word logistics, originally developed as military logistics, is one of three military terms that comprise the theory of war: strategies, tactics, and logistics. On the other hand, logistics, which supports daily life in cities, refers to a series of processes such as transportation, loading and unloading, and storage for the distribution of goods from producers to consumers. Physical distribution, which is a part of logistics, is divided into transportation links (transportation routes) and nodes for storage and processing (transportation hub facilities) and has five functions as described below.

1. Transport functions: transport, pickup, and delivery
   The transport function refers to the function of moving goods and commodities spatially from the place of production to the place of consumption using various means of transportation. Collecting goods from multiple production sites and transporting them to a specific location is called "cargo collection", transporting them to a relatively distant location is called "transportation", and delivering the items transported in bulk to individual consumption sites is called "delivery".
2. Cargo handling function: loading, unloading, inspection, and sorting

The cargo-handling function refers to the loading and unloading of goods from trucks and other transporters, as well as the inspection and sorting of goods. Moreover, operations such as the replacement and transshipment of goods that take place in facilities such as terminals and warehouses may also be treated as part of the cargo handling function.

3. Storage function: storage and deposit

The storage function refers to the retention of goods and supplies in a warehouse or other location for a certain period. If transportation is a spatial movement, storage is a time movement. Storage refers to the maintenance and management of goods over a long period of time, while deposit refers to the temporary retention of goods for a short period of time during the distribution process. It is an important function that adjusts the amount of goods transported in response to demand. ·

4. Distributive processing function: handing, processing, and assembling

The distributive processing function is to add value to products by processing and assembling them. Production processing is the processing of the product itself to make it value-added. It refers to the process of changing a commodity, for example, into a form that is more acceptable to the market, such as processing fish into sashimi. On the other hand, sales promotion processing is the process of putting a price tag on a product or combining it with other products to create a set product.

5. Packaging function: packaging and wrapping

The packaging function is to protect the goods with materials or store in containers in order to maintain or add value to the quality of goods and supplies. Industrial packaging is for maintaining the quality of goods and supplies, while commercial packaging is for adding value. The former is wrapped in cardboard to prevent the goods from being damaged, while the latter is wrapped in well-designed wrapping paper.

These five functions are brought together in an information system. Efficient management of logistics information (quantity, quality, operations) and commercial distribution information (order receipt and delivery, finance) will enable smooth distribution of goods. The supply chain, which is a series of processes from the procurement of raw materials to production, distribution and consumption, is currently attracting attention, and management of an optimal supply system for goods and commodities is called Supply Chain Management (SCM). The remarkable development of information technology in recent years supports this SCM.

8.1.2 Trade and physical distribution

To correctly interpret the place of logistics in city planning, it is necessary to understand the difference between "physical distribution" as it appears in logistics and "freight transport" dealt with in city planning. Kuse (1999) organizes them in terms of logistics and city planning as shown in Figure 8.1. Firstly, the transportation in city planning can be divided into two categories: person trip and freight transport. The traffic of goods is called freight transport or goods movement, which pay attention to the movement of goods as cargo, and mainly focuses on the two functions of transportation and cargo handling. To summarize, "distribution" is defined as the addition of "trade" to physical distribution. Logistics is defined as the addition of production (e.g., procurement) and consumption (e.g., recycling and disposal) to distribution.

The distribution from producers to consumers consists of two parts, commercial distribution (trade) and physical distribution,

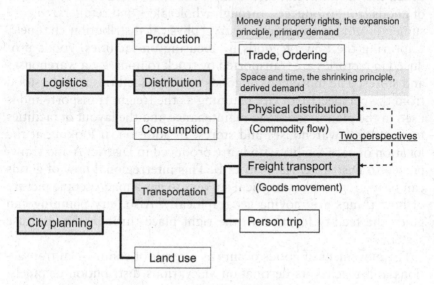

Figure 8.1 Physical distribution and freight transport.

Source: Modified from (Hirohito Kuse: *Value-Added Logistics*, Zeimukeiri Kyokai Co., LTD, 1999).

and which is extremely important to understand logistics. Commercial flow based on the principle of expansion is the transfer of ownership and money, and its profits are generated by selling to farther places, and at higher prices. On the other hand, physical distribution is about moving goods spatially and temporally (through storage), and it is based on the principle of shrinkage, which allows us to reduce expenditures by moving goods closer together, at lower cost, and in smaller quantities. From the aspects of demand, trade is the primary demand, and physical distribution can be considered as the derived demand. Therefore, the comprehension of the trade itself which generates goods movements is quite necessary to correctly understand the reality of logistics and to examine the efficiency of physical distribution.

8.1.3 City and distribution channel

The path in which goods flow from producer to consumer is called a "distribution channel". If the consumer likes the product and makes a purchase, the money is paid to the producer through the retailer. This sequence of goods ordering and receiving flows is called the "commercial distribution channel". Meanwhile, the flow of goods from producers, through wholesalers and retailers, to consumers becomes an inter-industry "physical distribution channel". Capturing the flow of goods by focusing on facilities, goods produced in factories are transported by truck to houses via warehouses and stores. The movement of goods between facilities is understood from the perspective of city planning as the freight transport, and is used in the planning of road maintenance and the layout of facilities for factories, warehouses and stores. In addition, if looking at the location of each facility, goods are produced in District A and transported to District C via District B. This interregional flow of goods can be aggregated in statistical surveys to get a macroscopic picture of how things are moving geographically. Also, city planning can guide the facility location to the right place through the land use system.

The movement of goods occurs as a result of commercial transactions and reaches its destination via various distribution channels. It is difficult to manage and guide the transportation volume itself directly, but city planning can indirectly affect the distribution channels by guiding the development of urban facilities related to logistics and optimizing land use.

R. Cox (1965) categorized urban industries into city forming industries and city serving industries. The former is an industry that encourages urban growth by allowing cities to trade with the outside world; wholesale commerce in the distribution industry falls under this category. The latter is an industry that provides a variety of services to urban residents; this includes retail, food and beverage, and service industries in the distribution industry. As the city expands, the area of commerce expands and wholesale commerce takes on an important role as a variety of goods are distributed. In recent years, with the advance of internationalization and informatization, wholesale commerce has separated into primary and secondary wholesalers. This has led to market competition, with many large retailers having wholesale functions and manufacturers encompassing wholesale and retail functions. In this multi-tiered or simplified commerce, the merchant who gains the upper hand in the distribution channel is called the Channel Captain or Channel Leader. Channel captains can not only determine the price of goods, but also change the distribution channel, such as changing trading partners. While human traffic can be assumed to take the shortest route to the destination, it is not always the shortest route in the case of transportation of goods. Observing the road traffic, one day the truck traffic may suddenly increase or decrease and these distribution channels need to be kept in mind in city planning.

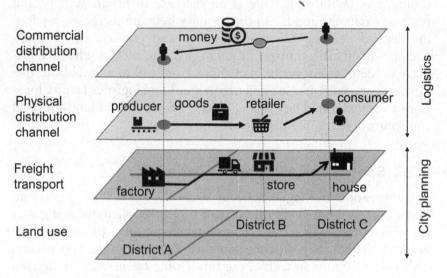

Figure 8.2 Distribution channels in city planning and logistics.

8.2 THE DIFFERENCE BETWEEN TRANSPORTATION AND PHYSICAL DISTRIBUTION

8.2.1 Primary and derived demand of transportation and physical distribution

Just as both of land use as a primary demand and transportation as a derived demand is inseparable in city planning, so is trade and physical distribution with both being closely related to each other in logistics. As land use changes, transportation also changes, so does the physical distribution if trade changes. For this reason, fully understanding the relationship between the two is necessary for planning and management.

The purpose of trade is to increase the profits associated with the transfer of ownership, while the purpose of land use is to improve people's quality of life (QoL) through the utilization of land. Examining trade and land use as primary demand is about considering how to stimulate economic activity and improve people's QoL. When examining physical distribution and transportation as a derived demand, how to move people and goods as efficiently as possible with the shortest travel distances and at the lowest cost should be considered.

Cities are places of vast consumption, and a huge number of goods are transported to support them. The exchange of these goods is trade, and the movement of goods is physical distribution. As the economy is revitalized, there is an increase in urban activity and trade to exchange goods. As the primary demand increases, so does the derived demand for transportation and physical distribution. Changes in efficient transportation in that era, such as when the primary mode of transportation changed from rail to automobile, also affects urban land use, leading urban areas and logistics facility locations to transform. Thus, trade, physical distribution, land use, and transportation are interdependent.

8.2.2 Spatial and temporal movement

Transportation and logistics occur when the benefits of spatial and temporal travel outweigh the costs of travel. People move and goods are transported when the benefits of arriving at a destination are generally higher than the time and cost of getting to the destination.

Considering human traffic, the travel time t generated by derived demands, such as commuting to work or school, should be as short as possible (see Figure 8.4). The concept of "the value of time" as elapsed time converted to monetary value (the opportunity cost of the time) is also extremely important in this case and is also used in the demand function in various transportation modeling. The value of time varies by personal attributes as well as by time and place. Although the exact calculation is extremely difficult, it can be generally calculated using the preference proximity method, which measures people's actual behavior, or using the income proximity method, which uses workers' wage rates. In addition, the determination of the amount of willingness to pay for reduced travel time forms the basis of people's decision-making process and is also an important theme in transportation economics, on which the viability of transportation projects and services depends.

In the case of logistics, on the other hand, the arrival time should not be early or late, and arriving at a predetermined time is important for just-in-time transportation. Therefore, in addition to the travel time t to transport the goods, there will be a time adjustment t' for long-term storage or short-term deposit. This time adjustment may be adjusted for the arrival of goods in a day, or it may be adjusted for weeks or even months, depending on demand. Thus, inventory management in logistics is a critical element of the supply chain and is carried out in accordance with the actual demand of the market sales trends. Inventory management is used to bridge the time gap between supply and consumption of goods, and can be divided by function: distribution inventory, which exists in the distribution channel from

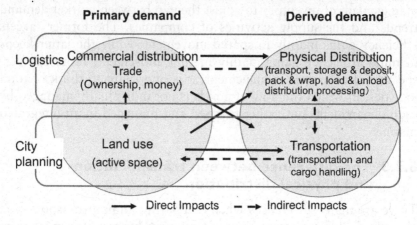

Figure 8.3 City planning and logistics.

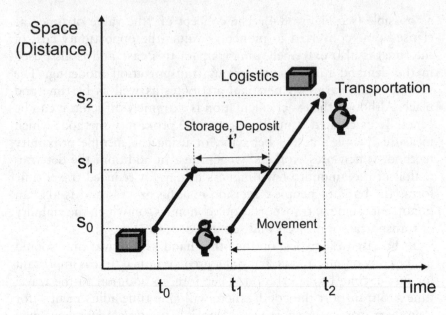

Figure 8.4 Spatial and time movements in transportation and logistics.

purchase to sale, and stockpile inventory, which is used to meet the demand. In recent years, how to reduce inventory has become an important strategy from a business management perspective.

The basis for transportation planning and logistics planning is to increase the benefits and reduce the costs of travel as much as possible. The major purpose of logistics planning is to achieve efficient logistics with the main objective of cost reduction and proper management of inventory to adjust the gap between market demand trends and the supply activities of companies. The former targets efficiency gains mainly in spatial movement, while the latter keeps in mind efficiency gains mainly in time movement. From the perspective of corporate management, it is important to think of total logistics, in which the flow of goods is not individually managed by each department, but rather grasped and managed in an integrated manner.

8.2.3 The difference between transportation and physical distribution

There are many differences when comparing human transportation and physical distribution. For example, in human transportation, movement itself may be the primary demand, such as a sightseeing trip or a stroll, but the movement itself is not the objective in physical distribution. Moreover, human transport does not change in weight or shape along the way, but physical distribution may change to a more value-added product during transportation, for example, through distributive processing.

Kuse (1999) summarizes the difference between the movement of people and objects as follows (see Table 8.1). First, for the movement of people, the unit of movement is generally invariant in terms of the number of people or the number of vehicles during a trip. For the movement of objects, however, there are various units such as weight (t), number, and size (m³) in physical distribution. Not only this, while the gender, age, or weight of people may be different, they are relatively similar as they are all people after all. Comparatively, physical distribution is much more diverse with over 3000 items in the shelves of a convenience store alone. The movement of people also does not change during the movement process, but physical distribution changes from raw materials to processed goods during the distribution process. In most cases, people can get in and out of traffic by themselves, but physical distribution requires people to load and unload cargo. Comparing

Table 8.1 The difference between person trip and freight transport

	Person trip	Freight transport
1) Unit	One (person); Invariable	Multiple (tons, m$_3$, pieces, dozen, etc.); Variable
2) Item	Human being only	Diverse
3) Process	No change	Changeable
4) Method	Self-movement possible	Requires loading and unloading
5) Purpose	Unity of purpose and action	Diverse travel purposes
6) Cycle	Complete in one day. Back to home	Various hours, days, months one-way traffic
7) Volume	Can be tracked by day and time	Change with time, day of the week, day, month and season
8) Commercial transactions	Mostly unrelated	Arises after or in anticipation of a commercial transaction

Source: Modified from (Hirohito Kuse: *Value-Added Logistics*, Zeimukeiri Kyokai Co., LTD, 1999).

the purpose of movement, person trip is consistent in purpose and behavior, while physical distribution is diverse depending on commercial transactions. While most cases of travel cycles are completed in one day for the majority of person trip and return to the origin (home), physical distribution is diverse and one-way in time. In terms of travel volume, person trip can be ascertained by day of the week and the time of day, but logistics varies by time, day of the week, day, and month.

This explanation highlights the diversity and variability of physical distribution compared to the movement of people. However, in some cases it can be more flexible. For example,. comfort on the move is a crucial factor for people, but goods do not complain about being placed in a crowded or dark area. As long as the quality of the product does not deteriorate, it can be moved for long periods of time. Therefore, there is a high degree of flexibility in storing goods on the move and adjusting the time to suit the needs of the distributor. The frequency of transport can also reduce by increasing loading efficiency. However, since there is a trade-off between efficiency and service in mass transportation and individual transportation, strategic decisions need to be made on a case-by-case basis.

8.3 CITY PLANNING AND PHYSICAL DISTRIBUTION

8.3.1 Hierarchy of transportation and physical distribution

Considering not only improvement of transportation routes, but also the location of transportation node facilities is crucial issues for physical distribution in terms of city planning. A large amount of goods can be transported in a city by connecting it to other areas by land, waterway, sea, and air. Goods are transported from other areas to truck terminals, ports, and airports, where they are also processed, packaged, and stored. The importance of nodal facilities tends to be underestimated due to invisibility by general users compared with visible movement on the roads, but the capacity of traffic nodal facilities would determine the volume of transportation in many cases. This is similar to the fact that traffic capacity depends on intersection capacity rather than road capacity. In any case, the capacity of the bottleneck has a significant impact on overall efficiency.

Therefore, it is crucial to determine the size and location of transportation node facilities in city planning. Especially in large cities, the development of ports and airports has a great influence on the future growth of the city. Large cities with ports and airports that function as hubs for wide-area logistics attract a large number of people and goods, which in turn leads to significant economic development.

Physical distribution has a hierarchy like human traffic. The upper part of the hierarchy is responsible for inter-city movement, and it is desirable to be able to move a large number of people and goods at high speed and low-cost using railways and highways. For intra-city movement, goods are sorted into medium-volume scales to meet demand and delivered at medium speed on the main road to the vicinity of the destination. For inner-district movement, the final step, the goods are delivered with small increments to the destination using manpower at low speeds. Transportation nodal facilities (nodes) are located at transit points that connect the hierarchical structure of pyramidal traffic routes (links) to efficiently redistribute people and goods.

8.3.2 Linking transportation and physical distribution planning

How can human traffic planning and physical distribution planning be interconnected in city planning? In an ideal situation, the flow of people and goods should be separated spatially and temporally. For

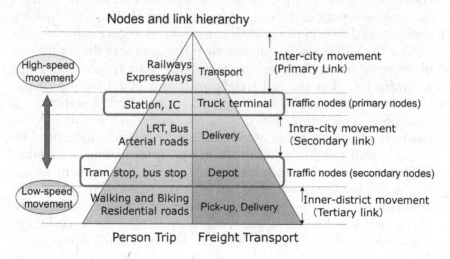

Figure 8.5 Hierarchy of person trip and freight transport.

example, just as a detached house has an entrance for people to enter and exit and a kitchen door for daily necessities, a hotel or large commercial facility has a front entrance for customers to enter and a backdoor for goods. In a large amusement park such as Disneyland, people move above ground while things move underground. Similarly, cities are expected to have separate travel routes for both. Therefore, the role of transportation routes should be determined at the stage of urban design. In addition, parking lot design is generally divided into general use and cargo handling use. Note that in cases where there are few users of the main roads, such as villages and small cities, human traffic and goods movement share the roads in terms of efficiency. Alternatively, freight and passenger consolidation policies are also being implemented for the sake of service efficiency in cases where demand is extremely low, such as in mountainous regions.

One of the ideal urban structures proposes a ladder structure with different primary lines of flow of human traffic and physical distribution. Public transportation between cities such as railways is the primary link for people, and highways are used as the primary link for physical distribution between cities. People and goods can share the roads on secondary links, and access routes to facilities may differ on tertiary links. The transportation nodal facility acts as the primary nodes, which connect the primary and secondary links. These are major railway stations for human traffic, or truck terminals, ports, and airports near highway ICs for physical distributions. The secondary and tertiary links are connected by tram stops and bus stops in the case of human traffic, and by depots in the case of physical distribution. The spatial conceptual diagram of the depot is shown in Figure 8.6.

Prior to the increase in automobile traffic, rail was the main form of intercity transportation (primary link). Rail freight still plays a certain role, but the rail freight yard as a nodal facility to the secondary link is increasingly being relocated from a station to a location at a certain distance from the station. This is due to the rise in land prices around stations as they became more central to the city, and freight facilities are moved to the suburbs to making better use of the land. Meanwhile, as the development of the inter-city highway network progressed, the focus of freight traffic gradually shifted from rail to road. Nowadays, as an efficient way of operating on highways, the technology of running several large trucks in a convoy by controlling the trucks is being investigated in an integrated manner through electronic linkage technology

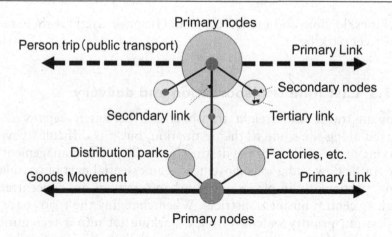

Figure 8.6 Ladder structure of transportation and physical distributions, primary, secondary, and tertiary traffic.

(vehicle-to-vehicle communication), which is also a way of using road space like a railway.

With secondary links within cities, human traffic and physical distribution basically share the roads. However, the delivery routes and delivery times of large freight vehicles within cities should be controlled, occasionally limiting them to small freight and van traffic. From the point of view of reducing road congestion and traffic safety within the city, facilities such as automobile terminals, warehouses, distribution and business parks should be located near major roads on the outskirts of the city. They should also be designed to not pass through the central city or residential areas. Truck inflow restrictions may also be enforced at midnight, early morning, and weekends.

Focusing on tertiary links (trip end traffic and pick-up & delivery), where the complexity of human traffic and physical distribution is likely to be a problem, the policy for the spatial and temporal use of road space is as follows.

- Access to the facility (front: people, rear: goods) Parked vehicles should be treated as goods.
- Road usage hours (weekday nighttime and holidays: people, weekday daytime: goods) Not in the city center, etc.

In addition, on-street parking should be avoided for both people and goods in the city center, but if there are no off-street parking facilities, loading and unloading should be prioritized for short-term on-street parking.

8.3.3 Last mile transportation and delivery

Human traffic and freight transport can be wisely separated or shared along the route to the destination, but it is difficult to avoid mixing in the vicinity of the destination facility. The management of the traffic flow and goods movement is an essential part of city planning, particularly in places where urban functions are concentrated such as central business districts. When changing the road space to pedestrian priority or converting a parking lot into a recreational plaza in order to create a pleasant city to walk in, it is not feasible as an urban policy if the freight transport in the area is not functioning properly.

Focusing on the movement of people and goods in the "last mile" to the final destination in transportation and physical distribution planning, it can be divided into three stages: access to the area, transfer points such as parking lots, bicycle parking lots, and cargo handling facilities, and approach to the facilities. The relationship to the arterial roads and proximity to the destination facility should be considered in order to ensure safe and smooth access routes. Door-to-door access to the facility may be required, or vehicles may be parked at fringe parking on the periphery of the city center for efficiency and city planning purposes. Secondly, unless visiting on foot, there is always a need for a ride-on and ride-off point for public transportation, a place to store vehicles for private transportation, and a loading area for freight transport in the district. This space is an essential part of operating a transportation system, but it can also be a problem in city planning. For example, it can cause traffic congestion due to illegal parking, illegal bicycle parking, or increased loading and unloading on the roadside. The priority of public transportation policy in the city center and the establishment of joint loading docks are proposed as one of the solutions to such urban problems. Finally, on the approach to the facility, human traffic such as people walking is unlikely to cause major problems, but in the case of physical distribution, there is the problem of horizontal delivery of goods and vertical delivery inside the facility. It is necessary to change the time of delivery so that a trolley loaded with goods does not have to come and go on a busy street with pedestrians, or if possible, delivery could be done through a back alley.

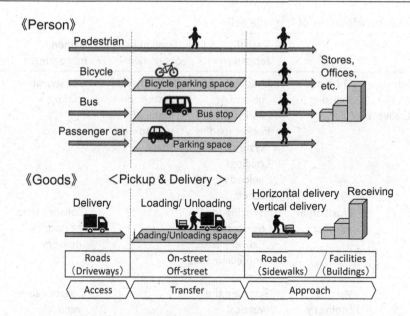

Figure 8.7 Last mile transportation and delivery.

Source: Modified from Tokyo Metropolitan Area Transportation Panning council. *A Guide to Pickup and Delivery Measure.* (Promotion of Logistics Measures together with City panning, 2015).

The delivery measures at the last mile that contributes to city planning can be broadly divided into three categories: spatial separation, temporal separation, and demand management (see Table 8.2). Spatial separation is desirable to plan the layout of roads, parking lots, or facilities when comprehensive infrastructure development can be carried out, such as the development of a wide area. In addition, environmental improvements can be made in stages by making it mandatory to provide parking and space for loading and unloading at the time of individual rebuilding. If there is a mix of human traffic and delivery in an existing urban area, there are ways to share space wisely using time. This includes a time sharing system on loading and unloading space and regional entry restrictions for cargo vehicles at various times of day. Alternatively, in order to reduce the number of freight vehicles entering the district, there are measures such as the development of depots for joint pickup and delivery, and the promotion of collaborative horizontal and vertical delivery. In any case, it is important to work with various related projects as part of urban development to solve the fundamental problem, rather than building the delivery system of the last mile in isolation.

Table 8.2 Measures of last mile delivery

		Spatial separation	Temporal separation	Demand management
Pick-up and Delivery	On-Road parking	Loading/ unloading parking Pocket loading system Loading/ unloading space	Time sharing system on loading/ unloading space	Collaborative pickup
	Horizontal delivery	Securing the horizontal delivery path and eliminating bumps		Collaborative horizontal delivery
	Vertical delivery	Securing the vertical conveyor channel		Collaborative vertical delivery
	Freight cars	Specification of freight car travel routes Regional entry restrictions for cargo vehicles, etc.		Collaborative pickup and delivery
Related projects	Infrastructure building	Land readjustment projects, urban redevelopment projects Development of traffic nodes, street maintenance		
	District transport measures	Congestion control in the district, barrier-free Construction of shopping malls, bicycle networks Measures to promote the use of public transportation Maintenance of private parking lots, public parking lots and pocket parks		

Source: Modified from (Tokyo Metropolitan Area Transportation Planning Council, *A Guide to Pickup & Delivery Measures: Promotion of Logistics Measures Together with City Planning*, 2015).

8.3.4 Transportation and physical distribution in the future

The relationship between transportation and physical distribution in the future will largely depend on the relationship between urban activity (land use) and commercial distribution, which are the primary demand. For example, when people want to buy something, they can choose to go out to the store to buy it themselves, or have the product itself delivered to their home. Choosing the former will generate shopping trips, while choosing the latter will lead to physical distribution from the store or warehouse to the home. Similarly, the choice of going to a restaurant or having food delivered also affects transportation and physical distribution. In recent years, with the advancement of information and communications technology (ICT), there are more activities that can be done at home, and the percentage of people going out has been decreasing year by year, while the number of home delivery services has been increasing. In the 10 years from 2008 to 2018, the going-out rate of residents in the Tokyo metropolitan area decreased by 10 percentage points, from 86% to 76%, and the number of trips per person per day decreased by 8% to 2.61 trips/person, while the number of parcel deliveries increased by 1.34 times (the overall figure for Japan). This trend was further accelerated by the new coronavirus pandemic in 2020.

In the future, as human traffic decreases and physical distribution relatively increases, urban facilities will need to be improved accordingly. There is an urgent need to take measures for the proper location of physical distribution facilities over a wide area, the formation of a distribution network, and last mile delivery coordinated with urban development. In addition, it is essential to respond to a society in which self-driving is becoming widespread. The challenge is to reorganize road space in a way that takes into account the balance between people and goods movements, such as eliminating illegal on-street parking that prevents autonomous vehicles from passing, developing cargo parking areas, and creating space for on-street cargo handling. Traditionally, two spaces, streets for movement and parking lots for storage, have supported the road transportation system, but with the spread of autonomous vehicles, it will no longer be necessary to provide parking lots at every destination, and the role of parking lots will change. Therefore, by converting a portion of the parking lot for passenger cars into a dedicated parking lot for cargo-handling vehicles, we can remove obstacles to automated driving caused by on-street cargo handling and achieve smooth and safe last-mile delivery. In the future, packages carried by autonomous vehicles will enter this special parking lot, and small automated delivery vehicles waiting there will drive on the sidewalk at low speed to deliver the packages to the facility's designated location.

Figure 8.8 Curbside designs of the future.

While the demand for parking will decrease, the demand for getting people on and off the streets for short periods of time and for the loading and unloading of goods will certainly increase. Therefore, in order to make effective use of the shoulders (curbside) of streets, it is recommended that flexible lanes be installed closer to the curbside to divide the use according to the time of day and demand. The image of the future street space is shown in Figure 8.8. The cross-section of the street will consist of walkable sidewalks, flexible lanes for various purposes such as getting in and out of the car, loading and unloading, environmentally friendly bicycle lanes, and a roadway for autonomous vehicles. This cross-sectional configuration is also a design that can flexibly respond to gradually changing demands for the future. There are many examples around the world, including the "parklet" in San Francisco, of using the section of the street between the sidewalk and the roadway as an open cafe or relaxation area. Spaces designed with the intention of improving the street space for pedestrians are called flexible zones (FZ), and there are increasing examples such as Castro Street (Mountain View, USA) in 1989. The proposed diagram shows a street configuration that emphasizes the proximity of sidewalks and flexible lanes. When separating automated driving from non-automated traffic (bicycles, pedestrians, etc.), the order is roadway, flexible lane, bicycle lane, and sidewalk. The appropriate street configuration will be selected by taking into account the spread of autonomous vehicles and the relationship with roadside land use. In any case, since streets within a city cannot be rebuilt all at once, this can be accomplished by repeated partial renewal in suitable locations.

REFERENCES

Shinya Abe, Shirou Uno (eds.). 1996. *Distribution and Cities in Contemporary Japan*, Tokyo: Yuhikaku Publishing Co., Ltd. (in Japanese).

Reavis Cox. 1965. *Distribution in a High-Level Economy*, Englewood Cliffs, NJ: Prentice Hall, Inc.

Sergio Jara-Diaz. 2007. *Transport Economic Theory*, Oxford: Elsevier Ltd.

Hirohito Kuse (ed.). 2014. *Introduction to Logistics*, Tokyo: Hakuto-Shobo Publishing Company (in Japanese).

Hirohito Kuse. 1999. *Value-Added Logistics*, Tokyo: Zeimukeiri Kyokai Co., Ltd. (in Japanese).

Hirohito Kuse, Yoji Takahashi, Dong-kun Oh. 1992. Basic Mechanism of Accumulation and Renewal of Physical Distribution Facilities in Tokyo, Selected Proceedings of the Sixth World Conference on Transport Research Society, Lyon, France.

Shinya Nakada, Masataka Hashimoto, Hideaki Kase (eds.). 2007. *An Introduction to Logistics*, Tokyo: Jikkyo Shuppan Co., Ltd. (in Japanese).

Hiroyuki Sasaki. 2014. Streets With "Flexible Zones" That Promote Pedestrian Usage of Parking Lanes, *Journal of Architecture and Planning*, 79(706):2661–2669 (in Japanese).

Study Group on Logistics of the CTI Engineering Co., Ltd. 2014. *Road Traffic Planning from the Point of View of Logistics: Dividing, Reducing and Switching Logistics*, Tokyo: Taisei-Shuppan Co., Ltd. (in Japanese).

APPENDIX

Tokyo Metropolitan Area Transportation Planning Council. 2015. *A Guide to Pickup & Delivery Measures: Promotion of Logistics Measures Together with City Planning* (in Japanese).

Chapter 9

City planning in cyberspace

9.1 USING ICT IN CITY PLANNING

9.1.1 Use of information and communication technology

The development of information and communication technology (ICT) is transforming urban life and business. In particular, smartphones have become an integral part of modern life, with one smartphone per person in general for people of all ages with its capability to respond to a variety of needs such as information acquisition, learning, entertainment, in addition to remote communication. For example, information that used to be obtained by traveling to a destination facility such as a conference room, cram school, movie theater, bookstore, or library is now available at home or even on the move, regardless of location or time. Outing activities such as shopping and eating out can also be substituted by the use of product and grocery delivery services. Recent surveys of transportation behavior have shown a downward trend in the rate of going out of the home for all generations, and this is especially true for younger people. This indicates that modern lifestyles are gradually changing. The pandemic of the COVID-19 in 2020 has led to a worldwide call to "stay home", causing a surge in demand for telework and even essential face-to-face get-togethers are beginning to become replaced by remote social gatherings.

The sharing economy is also evolving through the use of ICT. This includes car sharing, co-working, where offices and meeting rooms are shared, and vacation rentals for tourism. Particularly since the 2000s, some companies have emerged as providers of platforms that make it easy to rent, buy, and sell items on the Internet, and the market for these businesses has grown exponentially along with the rapid growth of businesses that utilize data.

The term cyberspace is defined here as a virtual space created by the data that users exchange on their computers. In contrast, physical space is defined as real space, and the relationship between cyber and physical space and city planning is explained in the following sections.

9.1.2 ICT-based transportation and logistics planning

Looking back at the modern history of transportation, if the 1820s–1920s was the era of the railroad, the 1920s–2020s was the era of the car, and the future 2020s and beyond will see a transition to an era in which "people will be the center of transportation". Cars gradually change from being owned to being used, freeing people from driving tasks while on the move. People are the protagonists in order to seamlessly use various transportation systems according to their characteristics. The technology that supports this is the network of cyberspace.

The various vehicles, which work to meet individual needs, will be able to coordinate information with each other to move more safely and efficiently. For example, road congestion information for the entire city is centralized in a road management center and reflected in the signal control. This can also contribute to the decentralization of road demand through users' information devices. In addition, joint transport and delivery and ride-sharing will be possible by matching underutilized vehicle information with user information.

The development of ICT has revitalized the sharing business, giving rise to ride-sharing and ride-hailing services in which private cars are dispatched to users at their request. The emergence of transportation network companies (TNCs), such as Uber and Lift in particular, has had a significant impact on the transportation environment. Although private car pick-up and drop-off services are not permitted in some countries, the simplicity and flexibility of the service over traditional public transportation has led to a surge in users and has changed the urban transportation environment, with a significant decrease in the number of existing taxis and public transportation users. On the other hand, changes in the transportation environment have naturally affected land use as well. For example, the widespread use of pick-up and drop-off services in conjunction with residential land development has caused a relative decline in property values near stations in some areas.

Furthermore, the proliferation of ICTs has prompted a major transformation in transport services which were previously provided independently. The concept of seamlessly linking various modes of transportation other than private cars through ICT to provide a single transportation service is called MaaS (Mobility as a Service). A unique feature of this service is that users can use their smartphones to search, book and pay for various transportation services, and can use the service as many times as they like within a certain area for a fixed price. Starting in 2016 in Helsinki (Finland), the concept of MaaS has spread across the world. At present, there are various levels of MaaS and Jana Sochor et al. (2017) categorize MaaS into five levels based on the integration and functionality of mobility services. They are Level 0 (no integration), Level 1 (integration of information), Level 2 (integration of booking and payment), Level 3 (integration of the service offer, including contracts and responsibilities), and Level 4 (integration of societal goals). At this point, almost all of the cases are still below Level 3, but it is expected that they will move toward Level 4 in the future. The impact of the introduction of MaaS on society and individuals is also diverse, and the Ministry of Internal Affairs and Communications in Japan has summarized examples of the impact as shown in Table 9.1.

Advances in ICT are also making significant changes in logistics. In recent years, there has been an increasing trend toward the supply chain management (SCM) to manage the optimal supply of goods and commodities throughout the entire supply chain from procurement of raw materials to consumption. In this context, ICT plays an important role in consolidating and managing physical distribution information (quantity, quality, operations) and commercial distribution information (order receipt and delivery, finance). This digitalization of transport and delivery can be divided into commercial flow, physical distribution and the infrastructure that supports them, as shown in Figure 9.1.

What is important in this context is the interrelationship between the order information system, which accelerates the flow of business, and the logistics information system, which improves the efficiency of logistics. Expanding commercial transactions is one of the business goals to be more profitable in business. The construction of the order information system makes it easier to receive and process orders from remote locations in a shorter period of time. This leads to an increase in logistics activities and an increase in both transport distance and volume. Meanwhile, as the logistics volume increases,

Table 9.1 Networked traffic information (MaaS)

The expected impact of MaaS on society and individuals		
Improving the sustainability of cities and regions	1) Reducing congestion in urban areas	Efficient travel by public transportation and new vehicles will reduce traffic congestion in cities by reducing the number of private car tips.
	2) Environmental Impact	The reduction of emissions from automobiles will reduce urban air pollution and greenhouse gas emissions. In addition, the reduction in the number of cars owned by private individuals will reduce the area of parking lots, making it possible to convert them into green spaces and other areas.
	3) Maintaining transportation in rural areas	The introduction of self-driving cars as service vehicles and the use of data to optimize the operation of buses and other vehicles will enable last-mile travel between stations and stops and destinations by people living in areas with limited transportation options.
Transportation Efficiency	4) Increased revenue for public transportation	Increased use of public transport, as seen in the Helsinki demonstration phase, could increase fare revenues and keep public funding through taxes low.
	5) Improving the efficiency of public transportation operations	Eliminating routes in areas where it is difficult to maintain the railroad and investing the funds to operate and maintain them in on-demand buses and self-driving cars would allow for more efficient operations.
Improving personal convenience	6) Integrated search, booking, boarding and payment	When transferring between multiple modes of transportation, a single service completes the search, booking, boarding, and payment of travel routes.
	7) Impact on household finances	Eliminating the burden of high private car maintenance costs will give you more room to devote to other expenses.
	8) Simplified travel expense reimbursement	This simplifies the expense reimbursement process for both companies and employees by allowing companies to pay their employees a flat rate for commuting allowances and making it easier to identify traffic routes other than the default commuting route.

Source: Modified from (Website of the Ministry of Internal Affairs and Communications in Japan. https://www.soumu.go.jp/menu_news/s-news/02tsushin02_04000045.html).

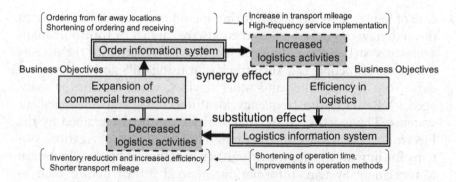

Figure 9.1 Substitution and synergy effect of logistics information systems.

Source: Modified from Hirohito Kuse (eds.): *Introduction to Logistics* (Hakuto-Shobo publishing company, 2014).

improving logistics efficiency becomes one of the management goals and a logistics information system is constructed. Through shorter logistics operations and improved operating methods, inventory reduction and transportation distances will be shortened and logistics activities will be reduced. In the former commercial flow, information and logistics activities are synergistic, while in the latter logistics, information and logistics activities are substitutable. In general, the increase or decrease in the amount of logistics within a city is determined by which of the two is greater.

Both human transportation and logistics systems can be categorized from five perspectives: destination, route, location, vehicle, and infrastructure, which can be organized as follows (see Figure 9.2). Destination information is mainly available on the Internet in the case of humans, and exchanged through an ordering system in the

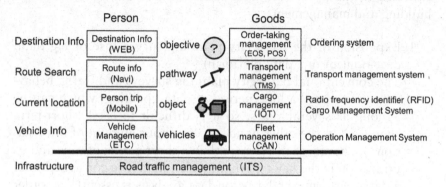

Figure 9.2 Smarter transportation and logistics.

case of logistics. In terms of information on routes, people are routed through navigation systems and goods are carried by transport management systems. Regarding the location information of the objects to be moved, people can keep track of it using GPS and base station data from smartphones and other devices, while logistics is managed by RFID (radio frequency identifier) and cargo management systems. The location of moving vehicles can be determined by the Electronic Toll Collection System (ETC) and car navigation systems for human traffic, while logistics is managed by the Operation Management System. Information on road infrastructure such as congestion is then provided by the Intelligent Transport Systems (ITS). At the present time, the systems have been constructed for different purposes, so mutual cooperation between them has not progressed, but it is expected cooperation and networking will progress in the future.

9.1.3 City planning with ICT

The progress of the information society has changed the conventional concept of space. Advances in communication technologies that transcend physical distance have directly connected diverse places and created new industries and communities. New planning concepts are needed in future city planning which are based on new lifestyles and businesses in order to respond to the paradigm shift in spatial concepts. In addition to the conventional city planning that targets physical space, we need to add the cyberspace embodied in ICT to comprehensively manage the city.

Considering the use of ICT technology based on traditional city planning techniques, several technological advancements are expected in the process of data collection and analysis, consensus building and management.

1. Expansion of the basic city planning survey (use of big data, development of a data platform)
   The conventional basic city planning surveys and traffic behavior surveys have a long period of data updating, which can be as long as several years, so it is difficult to catch short-term changes. On the other hand, an enormous amount of information can be stored in cyberspace in a short period of time and its utilization is the key to dynamic planning. The construction of a data platform as a basis for data analysis is essential in order to handle such big data in city planning.

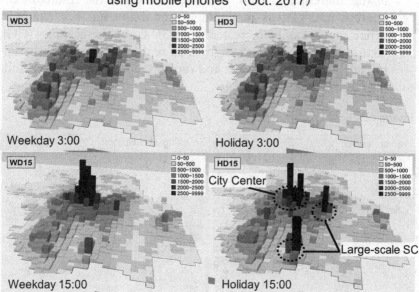

Figure 9.3 Short-term evaluation of urban structure using big data.

2. Evolution of analysis methods (use of artificial intelligence)

   In the field of statistical methods, new analytical methods are necessary to explore in addition to the conventional ones. Until now, the mainstream method of urban analysis has been to aggregate the collected data and analyze its trends. However, to cope with big data that is sequentially updated in an extremely short period of time, it is necessary to have a method that automatically updates the analysis results at every data acquisition. Therefore, the development of artificial intelligence (AI) such as "Reinforcement Learning", Bayesian theory as well as frequency theory is attracting attention in statistics.

3. Expansion of consensus-building methods (visualization of urban space, use of 3D-CG)

   In city planning, public involvement (PI) is important to actively provide information and communication to citizens and other stakeholders at every stage from the draft to the completion of a project. Face-to-face information exchange such as public relations meetings and workshops has been common until now, but in the future, it will be necessary to actively utilize social networking services (SNS) in cyberspace to exchange information.

This requires everyone to be able to easily understand the results of the analysis and plan, and visualization of the results of the analysis and plan is more important than ever. For example, urban spaces can be recreated using 3D computer graphics (CG), virtual reality (VR), and augmented reality (AR) to visually represent the planning process and future visions to aid understanding (see Figures 9.4 and 9.5).

Image of Ikebukuro city in 2050

Figure 9.4 Future Tokyo in 2050 with VR.

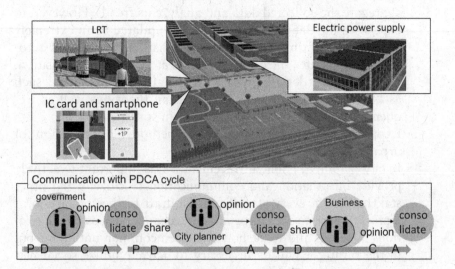

Figure 9.5 Visual communication with 3D-CG.

4. Management organization (a new public–private partnership)

The purpose of modern city planning is to "promote the sound development and orderly maintenance of the city", which has been carried out mainly by the government because of its high public interest. However, since cyberspace is mainly constructed by the activities of private companies and individuals, a new management organization of public–private partnerships are expected to play a certain role in order to integrate with traditional city planning. Here, the balance between efficiency and the public interest is important.

9.2 URBAN MODELS IN CYBERSPACE

9.2.1 Smart city

The Internet of Things (IoT), where things are connected to the Internet and exchange information with each other, is starting to be used in a variety of fields. The term "smart city" refers to a city or district in which various IoT sensors within a city gather a vast amount of information, and overall optimization is achieved using ICT. The number of cities that have adopted smart city policies began to increase from the early 2010s, and the movement toward smart cities is accelerating in many parts of the world.

The early days of smart cities have been driven by smart electric power systems. The main technology for this is the "smart grid", which is a system that controls the flow of electricity from both supply and demand perspectives to optimize the balance between them. Alternatively, "microgrids" are a power supply system that actively utilizes distributed power sources (solar, wind, biomass, etc.) within a specific region and utilizes information technology (IT). Particularly in the United States, there is a great need to rebuild the efficient operation of the power transmission and distribution networks against the backdrop of the power crisis, and plans for smart cities have begun with smart grids as their mainstay. Meanwhile, in Europe, smart city initiatives have been launched as part of the countermeasures against global warming, a concept that follows on from the eco-city, which aims to make cities more energy efficient and low-carbon. In contrast to the cases in the United States and Europe, which have mainly focused on existing cities, smart cities in emerging economies are often characterized by the development of new cities on vacant land, and their objectives are more inclusive, including the creation of jobs and new industries. The beginning of the smart city movement in Japan came in 2010 with the launch of "smart community" experiments in four locations across the country. The selection of towns and communities as sites for technology demonstration is unique compared to the US and Europe.

Smart cities around the world started out in individual fields, such as energy, but have gradually become cross-sectoral in nature, expanding from a specific area to a wide area involving entire cities. The EU's assessment of smart cities is based on the following six characteristics.

- Smart Economy: competitiveness
- Smart People: social and human capital
- Smart Governance: participation
- Smart Mobility: transport and ICT
- Smart Environment: natural resources
- Smart Living: quality of life

A variety of Key Performance Indicators (KPI) has been introduced and efforts are underway to leverage ICT and IoT to build a richer and more sustainable society.

The Japanese government proposed the Society 5.0, which is "A human-centered society that balances economic advancement with the resolution of social problems by a system that highly integrates

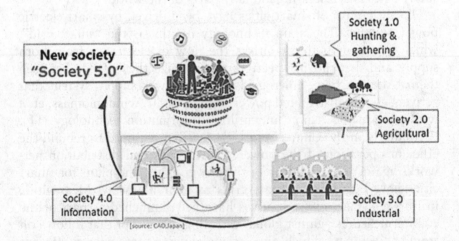

Figure 9.6 Society 5.0.

Source: Reprinted from Cabinet Office in Japan (https://www8.cao.go.jp/cstp/english/society5_0/index.html).

cyberspace and physical space". This new concept was proposed in the 5th Science and Technology Basic Plan as a future society that Japan should aspire to. It follows the hunting society (Society 1.0), agricultural society (Society 2.0), industrial society (Society 3.0), and information society (Society 4.0). In Society 5.0, vast amounts of information from sensors installed in physical space will be accumulated in cyberspace, and the big data will be analyzed by AI, and the analysis results will be fed back to life in physical space in various forms.

9.2.2 Smart cities and compact cities

While smart cities can contribute to solving various problems within cities, there are many challenges in terms of their compatibility with conventional city planning. Particular attention should be paid to the affinity with compact cities, which have been proposed as a sustainable urban model. For example, if self-driving cars become more prevalent as the technology of smart cities progresses, the transportation convenience of suburban areas is expected to improve dramatically, reducing the relative advantage of being located in the city center or near a train station. While aiming for urban compactness in physical space, we are trying to build a system that supports low-density, diffuse cities in cyberspace. In other words, it should be noted that there are significant differences between the two in terms of means and principles, while striving for the same sustainable city.

The difference between a compact city and a smart city can be simplified and summarized as shown in Table 9.2. While compact cities can be visualized because they mainly target physical space, smart cities are difficult to visualize because they are constructed in cyberspace. In contrast to compact cities, where public institutions

Table 9.2 The difference between a compact city and a smart city

Model	Compact city	Smart city
Object	Space	Information
Visibility	Visible	Invisible
Principle	Shrink	Expand
Method	Planning and management	Connected technology
Subject	Public	Private
Period	Long-term	Short-term

take the lead in planning and management over a long period of time, smart cities are led by private institutions and require short-term results through information connected technology. The biggest difference is that compact cities are based on the principle of spatial shrinking, whereas smart cities are based on the principle of information expansion. If the implementation of each policy is weak and the effect of policy implementation is partial, it will not be a major problem, but if the both policy is continued over a long period of time, it may create new issues in urban formation. There may be conflicting effects of expansion and contraction on the urban structure, and new disparities may emerge between places with high and low profitability. When aiming to integrate physical and cyberspace, it is necessary to comprehensively discuss on a regional basis how to integrate different urban models to form a better society.

9.3 PROPOSALS FOR A NEW URBAN MODEL THAT FUSES PHYSICAL AND CYBERSPACES

9.3.1 Hierarchy in physical and cyberspace

Services and systems that directly connect physical and cyberspaces have already been introduced in various fields such as smart grids, automated driving, and medical monitoring, and are collectively referred to as cyber-physical systems (CPS).

Advances in CPS in each field remap individual cyberspace, which in turn affects the arrangement of physical space as it exists in reality. In other words, changes in cyberspace will trigger changes in transportation and logistics systems, which in turn will change future land use. Therefore, focusing on the interrelationship between transportation and logistics, the components of physical and cyberspace is organized into four levels (see Figure 9.7).

1. Demand level (cyberspace): travel demand and purchasing demand
   This is the level of thought at which people make essential choices, such as whether to go to a destination or have goods come to them. The two can be an alternative relationship, influencing whether to go out or not. It depends on essential factors such as people's values and lifestyle. As a result of a person's choice behavior, when that information is collected by the Internet or sensed by the IoT, information about demand is manifested and appears in cyberspace.

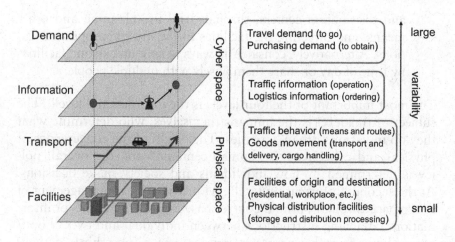

Figure 9.7 Physical and cyberspace (transportation and logistics).

2. Information level (cyberspace): traffic information and logistics information

This is a level of information about movement and transportation, including efficient routes and means of transportation. Information flies around in cyberspace, stored as big data and customized to the needs of the user. The system which centrally manages traffic and logistics information will contribute to the efficiency of traffic flow as well as to the improvement of loading rates and the consolidation of freight and passengers.

3. Activity level (physical space): traffic behavior and goods movement

This is the level of spatial and temporal separation and sharing of movement space. For example, it indicates choosing the means and routes to avoid congested times and areas, and efficiently transporting, delivering, and loading and unloading goods. When traffic action and goods flow overlap, the road space can be used effectively by sharing traffic and physical distribution well, by devising times for cargo handling priority and on-street parking areas.

4. Facility level (physical space): facilities of origin and destination and physical distribution facilities

This is a level that considers the proper placement and use of various facilities according to their functions. Preventing the mixing of facilities with low affinity, such as residential areas, factories and distribution centers, through land use planning and other means, and directing the facilities to the most appropriate locations. Alternatively, considering ways to split up existing facilities by time of day or share them wisely with multiple people.

The most important of the four levels is the top "demand level". The subject of this level is the humans themselves, who determine what they do in various urban activities. The origin of the convergence of physical and cyberspace is "human-centered" and the overall policy is determined by how individuals and society make decisions. At the second "information level", AI provides ideal management of cyberspace through optimal control based on vast amounts of information. Balancing instructions between individual and overall optimum allows for efficient use in physical space. The third "activity level" and the lowest "facility level" targets physical space, so smart sharing of limited time and space is the key to efficiency. This could include the installation of "flexible lanes" between the sidewalk and the roadway, where the use of the roadway can be varied to meet demand. While demand and information at the upper levels fluctuate and are fluid in the short term, the lower levels are more spatially constrained, and the layout of facilities is fixed in the medium to long term once it is decided.

9.3.2 Smart sharing city

The compact city which aims to consolidate urban functions in physical space, and the smart city which uses ICT in cyberspace to solve urban problems, were proposed under different concepts. Therefore, new concept that encompasses the two urban models is needed in order to integrate them. What compact cities and smart cities have in common is that they aim to be sustainable by "sharing space and information wisely".

A new concept, smart sharing city (SSC) is defined as "a city where individuals (or groups) share space and time wisely to maximize not only individual benefits but also social benefits". In other words, it is a city that efficiently and jointly uses non-operating assets to realize a sustainable society. Sharing is generally established when the benefits of the entity are enhanced by the sharing. Sharing is motivated by increased personal benefits, such as lower fares through carpooling, reduced labor through joint transport and delivery, and lower rents through room sharing, all of which increase personal benefits. On the other hand, maximizing individual benefits may not necessarily

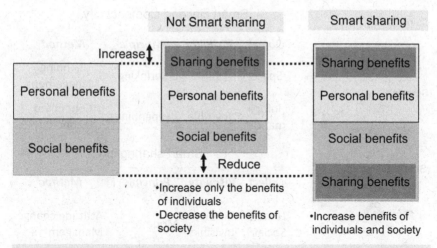

Smart Sharing City : "A city that efficiently and jointly uses non-operating assets to realize a sustainable society"

Figure 9.8 Concept of smart sharing city.

lead to maximizing social benefits. If everyone uses self-driving cars, traffic jams may occur and the impact on the environment can be significant. SSC is characterized by minimizing behavioral changes that do not increase social benefits, leading to smart sharing that benefits both individuals and society (see Figure 9.8).

Compared to the traditional urban model, if compact cities are based on the principle of spatial shrinking and smart cities are based on the principle of information expansion, then SSCs are based on the principle of social moderation. The method to achieve this goal is time and place sensitive management and individual attitude change. While the former intervenes exogenously to increase the benefits of society, the latter is a method of endogenously guiding individuals and society in the desired direction. In order to maintain public interest and efficiency, it is preferable for the organization to be run by a partnership of public and private entities (see Figure 9.9).

The development of the desired transportation and land use relationship achieved by the SSC is a challenge in how to share urban and mobile space wisely. Assuming an urban structure that shifts from high to low density from the city center to the suburbs, the following is an example of the desired relationship (see Figure 9.10). High-density downtown traffic will be centered on walking, with a multilayered public transportation system within walking distance. Within the city, transit-oriented development (TOD) will be

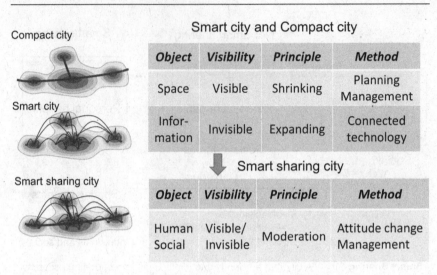

Figure 9.9 Features of smart sharing city.

implemented around the train station, creating an urban area within walking distance of concentric circles. The suburban areas are connected to the city center in a corridor shape along a fixed-route public transportation axis, while the low-density outer reaches are covered by demand-driven door-to-door traffic based on automated driving. Each transit system will be connected through ICT and AI will

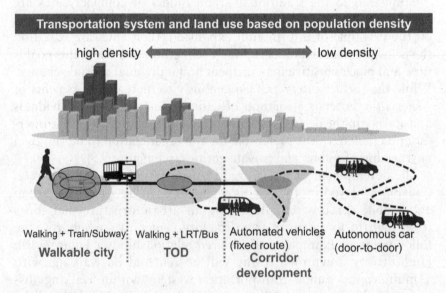

Figure 9.10 Land use and transportation in SSC.

support seamless travel for users while aiming to optimize the entire system. It is vital that the system in cyberspace supports efficient and comfortable urban activities while constantly monitoring the physical space.

In a democracy, the whole economy is made up by the accumulation of individual economic activities. The basis of economics is that resources are allocated efficiently and social surplus is maximized as a result of each individual's free benefit-maximizing actions. However, the market is distorted by the presence of externalities in the real world, thus creating the validity of market intervention by government. The externalities of increased environmental impact have given rise to compact city policy, which can be seen as a top-down policy decision that guides the overall optimization of cities. Smart cities, on the other hand, were created to make effective use of unused time and spatial resources by allowing different industries and business types to work together through ICT. It is a bottom-up policy because collaboration improves the benefits of the actors and this is the driving force behind the policy drive. If there were a market to merge the two policies, the two could be automatically coordinated within market principles, but no such market exists at the moment.

In economics, when external diseconomies occur, a way to internalize the external economy is the introduction of "Pigovian tax". This tax is intended to correct undesirable market outcomes (a market failure) by being set equal to the external marginal cost of negative externalities. As a countermeasure against global warming, an environmental tax based on CO_2 emissions is an example of a Pigovian tax. Although it is necessary for the government to obtain accurate information on external diseconomies at the time of taxation, it is extremely difficult to predict in advance what kind of external diseconomies will be generated by the effects of ICT-based information coordination. Meanwhile, "Coase theorem" states that external diseconomies can be eliminated if a place for negotiations between the various economic entities is prepared, even without government intervention. The idea is that if there is no cost to negotiate, as long as the benefits of negotiation exist, it will result in an appropriate allocation of resources and avoid market failure.

To support this idea, the essence of a SSC is to create a new market that connects the benefits of both. In other words, one of the ways to do this is to form a platform for maximizing both individual benefits and societal benefits. Such a merging measure is expected to correct the distortions in the planning theory of physical space and cyberspace.

9.3.3 Building a platform to support the city of the future

The structure of the platform which connects physical and cyberspace will be explained. An important role of the platform is the free sharing of data between each entity and the proposal of alternatives that increase the social benefits. First of all, we have to construct a database to manage various big data including real-time data according to the level of confidentiality of the information. For example, there are three levels of sharing: the public access level, which is open to everyone, the specific sharing level, which can only be shared by participating organizations that have accepted a user agreement, and the administrative retention level which is not shared with others. The other is the development of an AI-based reasoning engine. Based on the information in the database, it is to build an analysis system that can present alternatives that increase social benefits while maintaining individual benefits. Taking human mobility as an example, the system is such that when you enter the information about the destination, the AI will suggest multiple optimal routes and modes of transportation. The AI instantly judges the information on traffic congestion and public transport usage at that point in time and offers users lower fares for alternatives with high social benefits and higher fares for

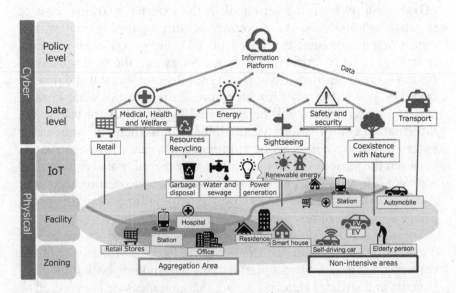

Figure 9.11 Integrated platform for smart sharing city.

alternatives with low social benefits. Users choose the best one from several alternatives to suit their needs. This creates a place where individual and social benefits are negotiated and automatically improves the social surplus. The privacy information including personal attributes is kept confidential in the system as the AI automatically processes the input information. If transportation costs are charged on a flat-rate basis (subscription system), the same effect can be expected by adding value-added points for actions that are in the public interest.

In addition, a mechanism for mutual evaluation by users is important to keep the platform in a healthy state. Lowering the rating if the transportation service used is less than satisfactory in relation to the rates quoted by the system can help to review the rates and remove poor quality service from the market. Alternatively, the information can be presented to users in terms of both price information and credit information (ranking) to encourage appropriate choices. As the application area of these platforms expands, they can take advantage of internal assistance schemes. Compensating from the more profitable routes to the less profitable routes will help maintain the entire transportation system.

On the other hand, there are many challenges in building the platform. For example, platform operators are given a lot of authority and accumulate a vast amount of information, which creates an "information asymmetry" between them and the participating companies. In addition, building and updating a platform to handle vast amounts of information takes a lot of time and significant expense. In terms of platform management, transparency and neutrality based on the premise of fairness are required, and at the same time, it is important to create sustainable mechanisms such as the creation of an autonomous decentralized platform.

Since the technological evolution of AI is faster than the speed of human technological evolution, it is assumed that it will eventually surpass human skills, and this point is called "Technological Singularity". It is envisioned that extremely advanced AI will eventually replace humanity as the basis for civilization's evolution. The platform proposed here does not manage people's urban life, but only presents an alternative to building a better society. It is important to state that it is the users themselves who choose the alternatives presented, and that the system is part of human-centered city planning.

The following dichotomies dealt with in this book are similar to two sides of the same coin and complementary to each other: Land

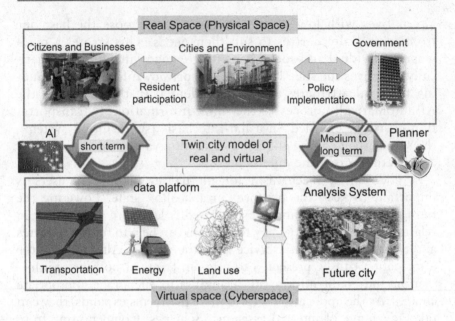

Figure 9.12 The image of twin cities model.

Use and Transportation, Primary Demand and Derived Demand, Human Transportation and Physical Distribution, Physical Space and Cyberspace, and Compact City and Smart City. Smart sharing is one of the guiding principles for skillfully reconciling the two and balancing the whole.

Eventually, all systems will be unified to manage Person Trip and Freight Transport more efficiently. When all self-driving vehicles are optimized at the management center, there will be no traffic jams or accidents between vehicles on the road. Energy consumption and environmental impact will also be controlled across the board and will be kept reasonable at all times. City planning is also a process of optimizing society towards a predetermined ideal image. In the distant future, AI may replace this extremely complex coordination in our society, but for the foreseeable future, it is an extremely big challenge for planners. Before the time comes when physical and cyberspaces completely merge, we need more than ever to discuss what is the desired output value of city planning in terms of using AI. This is because the aforementioned platforms also allow for coordination between existing entities, but the coordination between the present and future societies is extremely difficult. The children who will be born in the future should be the protagonists of future city planning, but there is no negotiating table because they are not present. It is hoped that those responsible for current city planning will take on a coordinating role on their behalf.

REFERENCES

Mitsuyuki Asano. 2014. *Transportation Spaces in Mature Cities: New Directions in Their Use and Renewal*, Tokyo: Gihodo Shuppan Co., Ltd. (in Japanese).

Yosuke Hidaka, Kazuhiko Makimura, Takekazu Inoue, Keizoh Inoue. 2018. *MaaS: Game Change for All Industries Beyond the Mobility Revolution*, Tokyo: Nikkei Business Publications, Inc. (in Japanese).

Hitachi and UTokyo Joint Research. 2018. *Society 5.0: Human-Centered Super-Smart Society*, Tokyo: Nikkei Publishing Inc. (in Japanese).

Japanese Cabinet Office.2016. Society 5.0. https://www8.cao.go.jp/cstp/english/society5_0/index.html (accessed August 15, 2020).

Hirohito Kuse (ed.). 2014. *Introduction to Logistics*, Tokyo: HAKUTO-SHOBO Publishing Company (in Japanese).

Ministry of Internal Affairs and Communications in Japan.2018. *Next-generation Transportation MaaS.* https://www.soumu.go.jp/menu_news/s-news/02tsushin02_04000045.html (accessed August 15, 2020) (in Japanese).

Akinori Morimoto. 2019. Towards the Integration of Compact Cities and Smart Cities, *The Journal of the Land Institute*, 27(2):10–15 (in Japanese).

Jana Sochor, Hans Arby, I.C. MariAnne Karlsson, Steven Sarasini. 2017. A Topological Approach to Mobility as a Service: A Proposed Tool for Understanding Requirements and Effects, and for Aiding the Integration of Societal Goals. *1st International Conference on Mobility as a Service*, Tampere, Finland.

The Centre of Regional Science (ed.). 2007. *Smart Cities – Ranking of European Medium-sized Cities*, Final Report, Vienna UT. http://www.smart-cities.eu/download/smart_cities_final_report.pdf (accessed August 15, 2020).

Katie Williams (ed.). 2005. *Spatial Planning, Urban Form and Sustainable Transport (Urban Planning and Environment)*, London: Taylor and Francis. Kindle.

Shinji Yamamura. 2014. *How Do You Create a Smart City?* NSRI Selections, Tokyo: KOUSAKUSHA Co., Ltd.

Chapter 10

Management for future city planning

10.1 DECISION-MAKING IN CITY PLANNING

10.1.1 Uncertainty in city planning

Even if an ideal vision of the future can be envisioned for the implementation of city planning, planners still need to identify various uncertainties and make the right decisions to address them. This section enumerates the types of uncertainty that exist in planning and presents a way of thinking about how to respond to them.

In decision-making, J. Friend and A. Hickling (1987) point out three uncertainties and explain the importance of "the strategic choice approach" in addressing them.

1. Uncertainties in the work environment (UE): uncertainty in research, analysis, prediction, etc.
2. Uncertainties about guiding values (UV): uncertainty about policies, objectives, priorities, etc.
3. Uncertainties about related choices (UR): uncertainty about communication, negotiation, consultation, etc.

Among other things, the route on coordination (UR) is the most important one in developing the planning concept as a strategic choice. Over time, rulers and inhabitants of cities change, and their decisions are greatly influenced by different environments and values. Uncertainty in the modern work environment and values is less abrupt and more predictable, but the predictability of uncertainty in decision-making is generally lower.

The economist Frank Knight has clearly classified future uncertainty as "risk" when it can be predicted probabilistically and "uncertainty" when it cannot be ascertained probabilistically. In city planning, policies are formulated, discussed and implemented by the decisions of various stakeholders. Such decisions are not subject to probabilistic laws, but rather are a realm of uncertainty as they fluctuate according to circumstances.

How do we respond to such an uncertain future in planning theory? There are two ways of thinking about this: a stacked attempt to reduce uncertainty as much as possible, and a way to derive what to do based on the assumption of an uncertain future. "Forecasting" is a way of thinking that uses current technology and knowledge to predict and respond to uncertain future situations as accurately as possible. This is a way of thinking that, according to future population estimates, urban populations will change and so transportation facilities should be developed accordingly. This method of thinking is widely used in various fields. On the other hand, "backcasting" is a way of thinking that involves envisioning what the future should look like and thinking about the path to achieve it. For example, in addressing global environmental issues, forecasting is a way of thinking about how many years later a percentage of carbon dioxide emissions can be reduced if current policies are implemented. Backcasting, on the other hand, discusses what needs to be done now, working backwards, in order to stop global warming by 2050 to reduce emissions by a certain percentage.

10.1.2 City planning and consensus building

The French historian F. Coulanges, in The Ancient Cities (1864), explains the process of the emergence of cities as "it took many years to constitute a city (civitas), whereas once this was agreed upon, the cities (urbs) were built all at once". In other words, cities (civitas) were built by religious and political consensus and therefore took an extremely long time, while cities (urbs) as a collection of buildings could be built in a short period of time because they were built after consensus. Here the decision-making of power and the construction of physical facilities are viewed separately, with the former being noted as more difficult than the latter. David Ewing (1969) also points out that many plans have failed because conventional planning has been too preoccupied with the numerical aspects and has missed the human aspects of planning, or because it has approached the human aspects in the wrong way.

How do we derive consensus for decision-making in city planning? From the perspective of behavioral economics, there is an approach from both market norms and social norms. Market norms here refer to "norms of monetary transactions" and social norms refer to "norms established by human relations". For example, consensus building for setbacks in road widening consists of financial compensation and cooperative action with the community and government. One of the reasons for the long under-start of road maintenance is that reasonable financial compensation has been prioritized and efforts to encourage cooperative behavior, such as prior negotiations, have been inadequate in many cases. Humans value social norms apart from calculating profit and loss, and once trust is broken, it's hard to get back on track. Behavioral economics explains that when market norms and social norms overlap, social norms are irreversible once they are defeated. This shows that once a system that was supported by morals has changed to a monetary system, traditional morals break down.

Similar phenomena are explained in terms of intrinsic and extrinsic motivation in social psychology, which is the basis of behavioral economics. Intrinsic motivation refers to motivation arising from interests, attitudes, and intrinsic values, while extrinsic motivation refers to motivation arising from external rewards and punishments. For example, when people voluntarily engage in environmentally conscious behavior, but are financially rewarded for that behavior, they lose their initial sense of value. It should be noted that extrinsic factors can undermine people's intrinsic motivation. Thus, in consensus building in city planning, it is essential to fully understand the social norms and intrinsic motivations of the people fostered in the target area and to take a careful approach from the beginning.

In addition, society is populated by people with diverse ideas, and each individual's rational choice may not coincide with the optimal choice for society as a whole. The conflict that arises in this process is called a "social dilemma". American psychologist Dawes, R. M. (1980), explained that three important ingredients for enhancing cooperation in social dilemma situations may be: knowledge, morality, and trust.

One of the most important aspects of city planning is the confidence-building. Various relationships of trust are established in cities where private goods such as houses and public goods such as roads coexist with each other, ranging from trust between business entities to trust between the government and residents. It is crucial to understand how trust is established because once broken, it is difficult to repair a relationship.

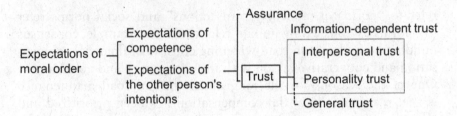

Figure 10.1 Conceptual classification of trust.

Source: Modified from Toshio Yamagishi. *The Structure of Trust- The Evolutionary Games of Mind and Society* (University of Tokyo Press, 1998).

According to Japanese social psychologist Yamagishi (1998), trust consists of expectations of competence and expectations of the other person's intentions. People can trust others when they can expect them to be highly competent and not to lie. Furthermore, expectations of intentions can be divided into assurance and trust. The former refers to the state of not lying because lying is damaging to themselves, while the latter refers to an estimate of goodness to the humanity of the other person. When explaining a city planning project to residents, it is necessary not only to clearly state that the implementation of the project is rational and will bring benefits to society, but also to make the residents understand that the planning entity is sincere and proposes the project from the same side as the residents.

The planners may want to build consensus while communicating the planners' intentions to the residents. This method of encouraging others to change their attitudes through persuasion is called "persuasive communication". However, it must be fully noted that "persuasion fails when the intent to persuade is felt".

When acting as an intermediary in such complex consensus building, it is helpful to understand "negotiation theory". The basis of negotiation is "mutual gains negotiation". This refers to a negotiation that is beneficial to all the parties involved and is also called a "win-win negotiation". At the root of negotiation theory is the exchange of goods that the other party values more than you and goods that you value more than the other party, thereby increasing the level of utility of both parties. In economics, this is called "Pareto improvement". What is required in city planning is not a win-lose negotiation where the other party is defeated with arguments, but rather a way to cooperate with each other to bring about a better situation. This means assuming an agreeable Zone of Possible Agreements (ZOPA),

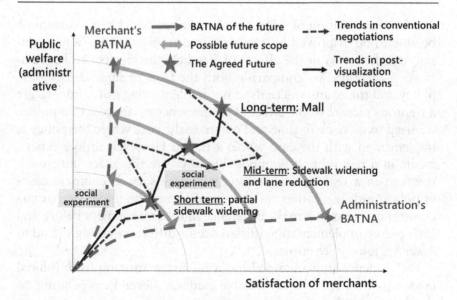

Figure 10.2 Decision-making using 3DVR.

and considering the Best Alternative to a Negotiated Agreement (BATNA) as a countermeasure to deal with the failure of the negotiations (see Figure 10.2).

For example, the introduction of a transit mall in an urban shopping district often causes opposition from shop owners in that area. That's because the majority of its current customers are car users, and it's concerned that curbing the influx of cars will lower its revenues. The government wishes to create a high-quality urban space for many users (public interest), and roadside merchants wish to secure revenue from the visitors. Although their values are different, what they both have in common is the revitalization of the city. A positive "trial and error" approach to urban development will help build consensus by using visualization of the near future, or a social experiment to help citizens understand the new city.

10.2 EVIDENCE-BASED POLICY MAKING

10.2.1 How to seek evidence in city planning

Evidence-based policy making (EBPM) will become increasingly important in city planning. Science is, according to Karl Raimund Popper (1902–1994), something that has falsifiability. In other words, the condition of science is that the proposed hypothesis must be able to be disproved by experiment and observation. When considering the terms of the effects of a policy, the effects can be accurately ascertained by comparing both the facts of implementing the policy and the counterfactuals of not implementing it. In order to get a rigorous picture of the effect of the presence or absence of a policy, we need to go back in time and compare the case where the policy is implemented with the case where it is not. However, unlike experiments in a test tube, it is impossible to create the exact same environment in a real city. Therefore, it is necessary to compare cases of policy implementation and non-implementation under similar circumstances, to measure the difference in the environment before and after policy implementation (difference-in-differences design), and to make various other efforts.

The evidence thus obtained has a hierarchy ranging from limited evidence (level 1) to highly reliable evidence (level 4), depending on the degree of reliability of the results.

Level 1: results from limited experiments and localized observations
   This is a study of trends in the data collected, and because limited experiments produce different results in different locations and subjects, it is positioned only as reference information. For example, results obtained by correlation analysis or simple regression analysis can show correlation, but causation is difficult to prove.
Level 2: results obtained from observational studies, natural experiments, and simulated experiments
   By making good use of real-world situations to separate the intervention group from the control group, which is not affected by the policy, a comparable group can be created to assess causality. However, social experiments that separate the intervention and control groups from each other can be difficult for ethical reasons, such as the effects of policies on safety. The use of accidental disasters and accidents to create comparable groups, for example, should be used in this process.
Level 3: randomized controlled trial (RCT)
   A randomized controlled trial (RCT) is a type of scientific (often medical) experiment that aims to reduce the source of a particular bias when testing the effectiveness of a new technique. By randomly assigning the subjects to the intervention and control groups, causal relationships can be

correctly assessed. Here, differences can be revealed by randomly assigning subjects to two or more groups, treating them separately, and comparing the measured responses. This is an ideal form of causal inference.

   Level 4: meta-analysis or systematic review

     The most reliable results can be obtained by synthesizing the results from multiple RCTs. A thorough survey of the relevant literature is required, and analysis should be conducted on data from high-quality studies such as RCTs, with as little bias in the data as possible.

Of course, a method that achieves high reliability is preferable, but this often requires a long verification period and a large amount of money, so the method is used differently depending on the object of analysis in practical terms. For example, the effectiveness of medication has a direct bearing on human lives and must be highly reliable in the field of medicine. However, because city planning deals with complex social phenomena, many variables have to be expressed in mathematics, and many assumptions and presumptions have to be made in order to actually solve this, which tends to reduce the accuracy of the details. This is because it is more important for policy evaluations to have the overall frame generally correct than to increase local accuracy. Thus, when deriving evidence in city planning, we have to take a bird's eye view of the structure of the problem and determine what methods and what logical reasoning to use to derive the evidence.

10.2.2 City planning and city analysis

Since city planning is aimed at the real world, the implementation of social experiments is limited, but various information about the city can be obtained through observations and surveys. Statistics is often used to analyze this obtained information. Statistics examines the regularity and nature of numbers from data and can be divided into following two categories; statistics based on frequency theory which represents uncertainty in data as a probability, and Bayesian statistics which represents uncertainty in a hypothesis as probability.

   Ronald A. Fisher (1890–1962) established modern inferential statistics, which is based on the premise of "frequentism". This concept defines "probability" as the frequency with which random events occur. It is a method of estimating the characteristics of a population by randomly selecting a sample from the population to be analyzed and is used in many areas of planning. It has been used in numerous social surveys, such as predicting the intentions of the entire urban population from a limited number of surveys.

On the other hand, Bayes' theorem discovered by Thomas Bayes (1702–1761) is a theorem about conditional probability and is unique in that it assumes some kind of probability in advance. The posterior probability is derived by multiplying the prior probability by the conditional probability. The probability of the cause is estimated from the result, which is being increasingly used with the spread of computers. The accumulation of sequential data improves prediction accuracy and is used in situations where practicality is more important than rigor, such as when sorting junk mail.

Logical reasoning in solving urban problems involves how to construct the causes and effects of various events. Taking land use and traffic as an example, the following approach is used when collecting examples of individual development and generated traffic to examine the general relationship between the amount of development and the amount of generated traffic. After statistical analysis of large data, we were able to derive, for example, the law that "large stores attract one visitor per day from one square meter of commercial floor space". This logical inference of collecting discrete data to derive a general law is called "induction". On the other hand, the general laws of development and generation can be used to predict the amount of traffic generated in an area. This is "deduction" which draws conclusions from logical reasoning based on facts and hypotheses. Using the previous rule, it is estimated that a commercial development of 10,000 square meters would attract 10,000 visitors. The inference of a hypothesis that best explains an individual event is called "abduction". This indicates that based on the fact that there were 10,000 visitors and one customer per square meter of commercial floor, the development is estimated to be the equivalent of 10,000 square meters of commercial floor space. Abduction starts from a set of observed facts and deduces to the most plausible explanation of those facts.

In summary, given assumption A and the rule "A then B" and the conclusion B, it can be said that deduction is to obtain the conclusion from the assumption and the rule, induction is to derive the rule from the assumption and the conclusion, and abduction is to infer the assumption from the rule and the conclusion. In city planning, when a general law cannot be found, induction is used to infer a law by collecting a lot of data and examples. Then, once the general laws are established, the premises and their laws can be used to find a special solution by deduction. Alternatively, abduction is used to investigate the cause of the problem and discover the problem (see Figure 10.3).

When analyzing cities, the solution of real-world urban problems can be achieved by defining a method of logical reasoning and utilizing mathematical methods such as statistics. It takes a different approach depending on the problem, but in general the process is as follows. First, a hypothesis is generated by carefully observing the real world and organizing the findings. If the hypothesis is substantiated by many trials, it is formulated and recognized as a general law. A mathematical model is a straightforward expression of that phenomenon through the power of mathematics. Furthermore, general laws can be used to derive individual concrete solutions. Also, hypotheses can be reviewed or new hypotheses can be drawn from general laws and results. The accumulation of such attempts creates a scientific basis for city planning. However, as Kozo Okudaira (1936–1979), a forerunner in the field of urban analysis in Japan, pointed out, "City planning consists of many parts that are difficult to understand or teach to others and a few parts that can be understood and explained objectively", the parts that can be proven are limited, so planners should always strive to use the scientific evidence with a humble attitude.

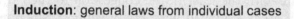

Induction: general laws from individual cases

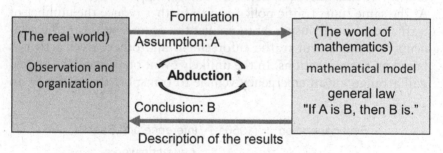

* Abduction: logical reasoning to derive a hypothesis that can best explain an individual event

Figure 10.3 The relationship between reality and mathematical models.

10.2.3 City planning and artificial intelligence

In recent years, artificial intelligence (AI) has been playing an active role in all aspects of people's lives. There are many examples of applications, from familiar household appliances to control technology in transportation and market research in cities. Although it seems like the latest technology, the term "artificial intelligence" was first used more than half a century ago. AI was first named at a conference held in Dartmouth, United States, in 1956, and AI research such as neural networks and machine learning became popular in the 1960s. After a brief lull, the commercialization of AI in the 1980s led to a resurgence of the AI boom, and expert systems were developed to reproduce the decision-making of experts. However, in the early 1990s, the boom rapidly cooled down as problems such as the maintenance and management of vast amounts of knowledge and its limitations were pointed out. If the 1960s was the first boom and the 1980s the second boom, the 2010s and beyond can be considered the third boom, and the current focus is on big data analysis using deep learning in conjunction with the Internet.

The technology to automatically collect various data in the city using IoT and analyze it in real time using AI is getting ready, and how to utilize it in urban planning has become an issue. AI is essential for real-time collaboration in platforms that connect physical and cyber space, and the fields in which it can be used will expand further in the future. For example, drivers can use AI to search for the best route with the least traffic congestion to their destination and travel in a car equipped with safety support functions by AI. At the same time, traffic police aim to further reduce the number of traffic accidents by using AI to predict the occurrence of traffic accidents and implement traffic enforcement and other street activities based on the predictions. In the unlikely event of an accident, the AI will arrange for an emergency vehicle to transport the victim to an

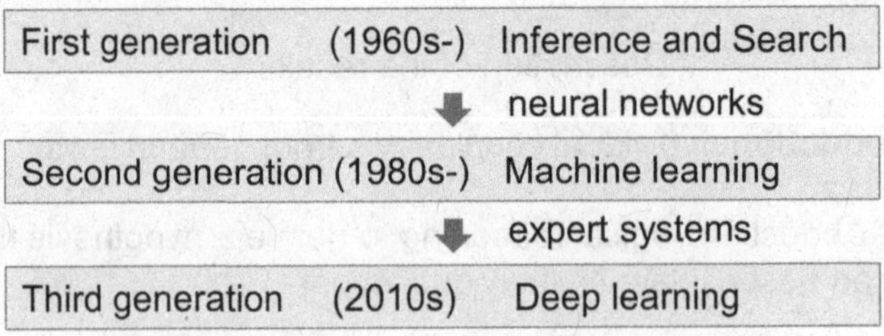

First generation	(1960s-)	Inference and Search

⬇ neural networks

Second generation	(1980s-)	Machine learning

⬇ expert systems

Third generation	(2010s)	Deep learning

Figure 10.4 History of the development of artificial intelligence.

acceptable hospital via the best route. Just as the technological innovation of AI has evolved with twists and turns over the past half century, AI technology will continue to advance, and one day there will
be a technological singularity where AI will surpass human capabili-
ties. Ray Kurzweil, an American futurist, proposed the concept in
his book published in 2005, and predicted that the Singularity will
arrive in 2045.

Also in 2013, Dr. Michael Osborne predicted that 47% of total US
employment is in the high risk category of computerization in 10–20
years due to the development of computers. Indeed, the evolution of
AI is likely to change many of the current occupational structures.
However, on the other hand, it is also expected to create new industries and occupations, and the debate continues to be a mixture of
expectation and uncertainty. Since city planning involves many elements of uncertainty, such as foreseeing the future and coordinating
various fields, it will be a long time in the future before planners
themselves are replaced by AI. What is important at present is to
think about for what purpose we develop AI and how we can give it
back to society. Planning for how to use AI to create a more affluent
society is a priority in city planning.

10.3 CITY PLANNING IN MODERATION

10.3.1 City planning entity

Considering planning in a broad sense, there are numerous different
entities that plan cities. City planning as an administrative plan has
been offering regulatory guidance and developmental activities for the
overall efficiency and public benefit of the city. If this is a top-down
type of city planning, there is also a bottom-up type of city planning
in which residents take the lead in improving the local environment.

In Japan, "community development" was born around the 1960s
as a citizen's activity and took root as an autonomous and continuous environmental improvement movement by citizens. To describe
it in a binary opposition, there is a "government-led city planning"
and a "citizen-driven city development". The former is mainly
used for public and professional planning actions and projects,
while the latter is used for localized, community-oriented activities.
While the former designs a city from a macro perspective and considers the balance of details, the latter considers the city to be constructed
by building up from the micro perspective of the neighborhood.

As is self-evident, creating a better city requires a two-sided approach. This is because different components of a city have different formation processes. When considered in terms of land use and transportation, which is the primary focus of this book, land use is often formed in a bottom-up fashion, while transportation is more compatible with a top-down approach. Since land use begins with the individual optimization of the development entity, the first priority is to improve the environment at the individual development unit. It starts with improving the immediate environment, and then it spills over into the surrounding area and makes the whole thing better. On the other hand, a broad-based perspective is essential for transportation planning, as it aims to facilitate the smooth movement between multiple points. This is because even if it is better in one place, it is useless if it does not function as a network. Therefore, the importance of hierarchy and networking is emphasized. In other words, there is no established theory that city planning can be achieved by a one-way approach only.

Even within the same field of land use or transportation, the idea of integrating the macro and the micro, the inside and the outside has been proposed. For example, Haruhiko Goto (2007) described community development in three models: "Exogenous" which is planned and organized from the outside; "Endogenous" which is planned and organized from the inside, and "Neo-endogenous" which is a model with both aspects. In contrast to developments from outside the region (exogenous), it has been pointed out that development from within the region (endogenous) should be promoted from the perspective of community building. However, he stressed the importance of developing autonomously, incorporating the best aspects of both in cooperation and collaboration with other cities and regions, rather than a model of development confined to the region.

Here, a fusion of an efficient hierarchical tree type and a flexible, crosscutting rhizome type was also proposed in the transportation network. It was also noted the importance of combining transportation modes for overall efficiency, such as public transit and cars, with diverse and flexible pedestrian-first transportation spaces. The "parklet", created as a space for people from the roadway, can be seen in recent road improvements and is a bottom-up approach.

10.3.2 The planner's philosophy

Even if scientific evidence is available, planners will always be forced to make some kind of decision when there are conflicting opinions on the merits of city planning. In order to deal with this situation, there needs to be a value standard or ideology on which to base decisions. It often depends on the discipline the planner has primarily studied and the position he or she is in when making decisions. When referring to the philosophy of city planning, it varies from planner to planner and therefore is not necessarily a unified view. It is also extremely difficult to extrapolate the thoughts of the planners. Taking these assumptions into account, we can look for clues to decision-making by inferring the planners' thinking from previous literature. For example, Ishikawa Hideaki (1893–1955), who is called the father of modern city planning in Japan, had the following words to say about city planning: "Love for society, this is called city planning".

It is not possible to specify the philosophy of city planning here, but I would like to introduce the word "moderation" as one of the important principles. Moderation means that the way of thinking and action is not biased towards one position, but is neutral. Moderation has been regarded since ancient times as one of the most important human virtues in both the East and West. Practicing moderation as an individual is also in line with the Buddhist teaching of "knowing sufficiency". There are limits to material wealth, and somewhere along the way people need to find spiritual satisfaction and live wisely in the world. Lao Tzu, a thinker of the Spring and Autumn period in China, taught in the Lao Tzu Moral Sutra that "those who know enough will be rich". The word "moderation" was first mentioned in the Analects, and Confucius (BC551–BC479) of the Spring and Autumn period wrote in the Analects that he praised "the doctrine of the moderation is one of the best virtue". Since then, it has been respected as a central concept in the Confucian tradition. The Doctrine of the Mean, now widely known as one of the four Chinese classic texts "Si Sue" is said to have been authored by Confucius' grandson Zi-si (BC481–BC402), and has been handed down as one of the pieces in the Book of Rites. In this context, he emphasizes that human beings can live in harmony with the universe by acting in moderation.

On the other hand, the importance of the idea of moderation has long been taught in the field of philosophy. Aristotle (BC384–BC322), who was called the founder of all sciences, believed that human happiness is "to strive throughout one's life to realize the virtue (arete) that is unique to human beings", and proposed the concept of "moderation" as the standard of virtue at that time. Moderation here indicated a state of no excess or deficiency, and encouraged people to live a proper life that was neither reckless nor cowardly, neither arrogant nor mean-spirited.

Moderation in city planning also means considering the harmony of the three elements that make up a city: time, space, and people. All three Chinese characters contain the word "between", implying that they exist between two events. The essence of city planning lies in anticipating the time axis, considering the balance of space, and coordinating people with people.

1. Time: generally speaking, city planning is a long-term plan, but daily phenomena are repeated in the short term. The results of decisions and actions made by individuals and businesses in a short period of time accumulate and affect the city over a long period of time. How do planners view the concept of time and how do they coordinate it? Coordination between current and future generations is also important in this process. One of the roles of planners is to bridge the gap between the very long term future and the present, especially since the economic principles of markets work in the short term.

2. Space: plans are often spatially limited by the target area. District planning is initiated by the residents of the district and city planning is driven by the residents of the city. However, the effects and impacts of the plan will spill over outside the target area as well. The development of one district can cause the decline of an adjacent district. How do the planners view the space targeted in the plan and how do they balance it with other areas? Planners should always try to coordinate not only the allocation of space in the target area, but also with space outside the area.

3. People: there is a diversity of people in cities. The purpose of city planning is for men and women, young and old, rich and poor, residents and visitors, able-bodied and physically challenged, to help each other and to respect each other's human rights. Planners must not only coordinate those who benefit and those who lose, but also take into account the diverse group of people. In general, the goal of city planning is to increase people's well-being, but well-being is not an absolute standard, but a relative one. Since people perceive their well-being by comparing themselves to the well-being of those around them, it is also important to reduce the disparity between people in order to increase social well-being.

The modern planner is only one of the guides who leads the city in a better direction, not the ruler who creates the city and society.

Planners are simply acting as a relay to pass on to future generations the cities that their predecessors in the past planned and continued to fix to fit the times. In that sense, city planning can be seen as a "technique of foresight and coordination" that makes an effort to see into the invisible future by making use of modern science and knowledge, and to carefully coordinate the various parties and matters involved.

10.3.3 Goals in the plan

What is the ideal form in city planning? It may be analogous to the imaginary ideal form "Idea" espoused by the ancient Greek philosopher Plato (BC427–BC347). There are various objections to this idea. For example, Aristotle pointed out that hypotheticals that cannot be verified in reality should not be the standing point of thought. The same could be said for city planning. City planning is also about building up realistic solutions to challenges that vary from time to time and region to region.

As long as city planning is planning for the future, it is important to make predictions about what the future holds. Alan Curtis Kay (1940–), the computer scientist who created the personal computer, noted that "The best way to predict the future is to invent it". We need to have a perspective that does not sit and wait for the future to come, but rather actively interferes in that future and creates it. Kevin Lynch, in his book "What Time is This Place?" (1972), also states that it makes sense to control the spatial present environment, act for the purposes of the near future, expand the scope of the open future, explore new possibilities, and maintain the ability to respond to change.

The pursuit of ideals in city planning is also the pursuit of increasing people's well-being. Subjective well-being is related to three things: economic and social status, physical and mental health, and connections to people and society. It is important to have a stable life with income and work, to be mentally and physically healthy, to build good relationships and to be connected to family and community. In other words, people's sense of well-being can be elevated and a happy society can be achieved by implementing a plan to enhance these three perspectives. On the other hand, what would happen to society if these three perspectives were taken to the extreme? We can imagine a society where people would all be very rich, close to immortality, and only connected to the people they love. In that society, much will be consumed, hastening the depletion of resources, causing an imbalance between generations and slowing down the generational cycle, which may lead to a loss of diversity and an increase in interspecies conflict.

There are limits to any policy and it is unclear whether the society beyond the policy is really desirable. The sooner we move forward with policies for a smart city, the sooner we may see a society where AI will oversee everything beyond that. If a compact city policy is successful, the city may become more attractive and attract more people, resulting in an expansion of the city. In the end, it may not solve all of the city's problems, it may just change the problem. City planning is a human-oriented discipline, and the more comfortable people become, the less they use their natural abilities, and the weaker the human functions may become, both physically and mentally. After all, the goal of city planning may not be to chase the maximization or minimization of things, but to aim for a point somewhere in the "moderation".

REFERENCES

Civil Engineering, JSCE. 2004. *Consensus-Building: The Dilemma of Agreeing in Principle but Disagreeing on the Details*, Tokyo: Japan Society of Civil Engineers (in Japanese).

Numa Denis Fustel De Coulanges. 2001. *The Ancient City: A Study on the Religion, Laws, and Institutions of Greece and Rome*, Ontario: Batoche Books Kitchener.

Robyn M. Dawes. 1980. Social Dilemmas, *Annual Review of Psychology*, 31:169–193.

David W. Ewing. 1969. *The Human Side of Planning: Tool or Tyrant?*, New York: Macmillan Company.

Carl Benedikt Frey, Michael A. Osborne. 2013. The Future of Employment: How Susceptible Are Jobs to Computerisation? *Technological Forecasting and Social Change*, 114:254–280.

John Friend, Allen Hickling.1987. *Planning Under Pressure: The Strategic Choice Approach*, Oxford: Pergamon Press.

Hiroo Fujita. 1993. *The Logic of the City: Why Power Needs the City*, Tokyo: Chuko Shinsho 1151 (in Japanese).

Haruhiko Goto. 2007. *Community Landscape Design*, Kyoto: Gakugei Shuppansha (in Japanese).

Hitachi and UTokyo Joint Research. 2018. *Society 5.0: Human-Centered Super-Smart Society*, Tokyo: Nikkei Publishing, Inc. (in Japanese).

Frank H. Knight. 2006. *Risk, Uncertainty and Profit* (Dover Books on History, Political and Social Science), New York: Dover Publications. Kindle.

Osamu Kurita. 2004. *Urban-Model Reader*, Tokyo: Kyoritsu Shuppan Co., Ltd. (in Japanese).

Ray Kurzweil. 2005. *The Singularity Is Near: When Humans Transcend Biology*, New York: The Viking Press.

Kevin Lynch. 1972. *What Time Is This Place?*, Cambridge: MIT Press.

Masahiro Matsuura. 2010. *Practice! Negotiation Theory: How to Build Consensus*, Tokyo: Chikuma Shinsho (in Japanese).

Akinori Morimoto. 2019. Smart Cities and Future City Planning, *The Ashigin Economic Monthly Report*, 121:8–13.

Yuki Morishige, Akinori Morimoto, Koki Takayama. 2018. A Proposal of Consensus Building Method Through the Visualization About Road Space Redistribution, *Journal of the City Planning Institute of Japan*, 53(3):1370–1376 (in Japanese).

Hisao Uchiyama, Yoh Sasaki. 2015. *Landscape and Design: The Basics of Civil Engineering from Scratch*, Tokyo: Ohmsha, Ltd (in Japanese).

Toshio Yamagishi. 1998. *The Structure of Trust – The Evolutionary Games of Mind and Society*, University of Tokyo Press (in Japanese).

Index

3D-CG models 142
3E measures 97
4M analysis 95

abduction 194, 195
adjustment and anchoring 85
advanced transport 125
agglomeration economics 8
aggregate model 83, 141
agricultural villages 1
Alonzo's bid rent 9
Amber Road 12
Amtrak 31
artificial intelligence 171, 196
augmented reality 143, 172
Automated Guideway Transit 125
Automated People Mover 125
automobile-dependent cities 59, 130
autonomous cars 128, 130, 137, 138
availability bias 85

B plan 67
backcasting 188
Baroque 7, 11, 15
barrier-free 113
Bayes' theorem 194
Bayesian theory 171
BCR 70
beneficiary burden system 11
Best Alternative to a Negotiated Agreement 191

bid rent theory 22
big data 82, 92, 170, 171, 175, 177, 182, 196
Bogota 63
bottleneck 89
British Railways 30
BRT 63, 88, 109, 125–131, 137
Brundtland Report 24, 47
Buchanan Report 28
bullet train 31, 79, 124
bypass road 57, 58, 93

capital gain 11
car sharing 61, 165
cargo parking areas 161
Chang'an 5, 6
channel captain 149
Cheonggyecheon 16
City Planning Area 43
Club of Rome 24
Coase theorem 181
commercial distribution 146–148, 160, 167
community development 107, 112, 118, 119, 197
community-based tourism 117
compact city 24, 25, 44, 47, 48, 110, 131, 175, 178, 181, 202
comprehensive plan 67, 108, 109
computable urban economic model 141
computer graphics 172

concentric ring model 22
consensus building 65, 134–137,
    142, 170, 188–190
Contingent Valuation Method 81
Copenhagen 16
COVID-19 111, 117, 165
co-working 165
curbside 138, 162
Curitiba 63, 109, 126
cyber-physical systems 176
cyberspace 166, 170–178, 181–184

data platform 170
deduction 194, 195
deep learning 196
derived demand 79–82, 85,
    102–106, 114–116, 148, 150
difference-in-differences design 192
disaggregate behavioral model 141
disaggregate travel demand model
    84
distribution channel 148, 149, 151
drones 128

eco-city 173
economic simulation 141
eco-tourism 117
Edmonton 126
elasticity 103
electric bicycles 127
electric kick scooters 127
electric vehicles 124, 127, 133
electronic linkage technology 156
Electronic Toll Collection System
    93, 170
environmental simulation 142
equal travel times principle 84
evidence-based policy making 191
expert systems 196

F plan 67
falsifiability 191
FAR 69–74
flexible lanes 162, 178
flexible zones 162
forecasting 188
four-step travel demand model 83
free ride 66

Freiburg 109
freight transport 147, 148, 158
frequency theory 171, 193
frequentism 193
fringe parking 55, 158
fuel cell vehicles 127

Garden City 19, 20, 33
general plan 67
generalization cost 82
GPS 81, 82, 170
Great Kanto Earthquake 66, 68
grid pattern 3, 5, 6

Heiankyo 5, 6
Heijokyo 5, 6
Helsinki 167
heuristic bias 85
hierarchy of urban transportation
    129
highway traffic capacity 89
horse-drawn carriage 15
Hotelling's model 9
Houston 63

IC 53, 54, 94
ICT 130, 161, 165–167, 170, 173,
    174, 178–181
incentive zoning 74
induced traffic 63
induction 194
information technology 72, 127,
    147, 173
integrated land use- transport
    model 141
Internet of Things 173
ITS 127, 170

Japanese National Railways 32, 33
joint transport and delivery 166,
    178

key performance indicators 174
Kyoto 6

land economics 58
land readjustment project 11, 54,
    68, 69, 72

land use simulation 140
Land Use Zones 43
landscape simulation 142
last mile 158, 159, 161
life cycle assessment 142
lifestyle-related diseases 113
local plan 67
location efficiency mortgage
   system 73
location management 73
location normalization plan 44,
   136
logistics 5, 145–155, 167–170, 176,
   177
logistics information system 167,
   169
LRT 88, 110, 125, 126, 129–133,
   137, 142

MaaS 167
machine learning 196
macro traffic simulation 91
management cycle 75, 76
Marunouchi 16
mass tourism 117
master plan 50, 67, 72, 88, 108,
   109, 131
megalopolis 4
metabolic equivalent 113
micro mobility 127
microgrids 173
micro-mobility 75, 130
micro-traffic simulation 91
Minneapolis 137
mixed land use 74
mobility management 89
mode choice 83
moderation 179, 199, 202
Mohenjo-daro 5
motorization 15, 26, 27, 32, 36,
   55, 59, 66, 107
Mountain View 162
multiple nuclei model 23
municipal master plan 67
mutual gains negotiation 190

NACTO 63
negotiation theory 190

neighborhood unit 20, 27, 28
network-type compact city 46, 47
neural networks 196
new transit system 125
New Urbanism 25, 35
New York 16, 30
non-excludable 66

OD traffic volume 81, 141
on-street cargo handling 161
operation management system 170
order information system 167

pareto improvement 190
Paris 11, 111
park-and-ride 52, 123
parklet 162, 198
PDCA 75
people mover 124
person trip survey 81, 82
personal mobility 127, 130, 132,
   133
persuasive communication 190
physical distribution 145, 155
physical planning 62
physical space 166, 170, 175–178,
   181
Pigovian tax 181
PLU 67
poisson distribution 97
population density 46, 63, 64, 113
Portland 109
primary demand 79, 102–106,
   113–118, 148, 150, 152, 161
principle of correspondence 69
probe car investigation 82
probe person survey 82
public health 110, 111
public image 21
public involvement 171
pyramids 12

quasi-public goods 12
queue length 90, 91
Q–V curve 89

Radburn system 27
radio frequency identifier 170

railway development model 33
randomized controlled trial 192
reinforcement learning 171
Renaissance 7, 15
representativeness bias 85
returning development profits 11,
    31, 66
revealed preference surveys 81
rhizome 47–51, 198
ride-hailing 166
ride-sharing 166
ring road 54–57
road pricing 61
road safety 80, 97
Roman road 13
Rome 6, 13
route assignment 83

San Francisco 162
SCOT 67
SDGs 99
sector model 22
shared bicycles 127
shared space 29
sharing economy 165
Silk Road 13
smart city 173–175, 178, 202
smart community 174
smart grid 173, 176
smart sharing city 178
social dilemma 189
social networking services 171
social planning 62
Society 5.0, 174, 175
space mean speed 89
spatial interaction model 141
stagecoaches 30
stated preference surveys 81
status quo bias 85
steam locomotive 14, 30, 79
Strasbourg 109
strategic choice approach 187
structure plan 67
subscription system 183
suicide 113
supply chain 146, 151
supply chain management 147, 167
sustainable city 25, 44, 48, 73, 175

sustainable development 24, 61
sustainable tourism 117

TDM 88
Technological Singularity 183,
    197
teleportation 79
TOD 35–37, 47, 52, 53, 66, 73,
    74, 131, 179
Tokyo 33, 35, 63, 65, 111, 161
tourism 115–117
tourism-based city planning 118,
    119, 120
traffic assessment 71, 72, 92
traffic calming 29
traffic cell 49
traffic congestion 89, 90, 93
traffic density 64, 89
traffic engineering 80
traffic flow rate 89, 90
traffic impact assessment 71
traffic impact fee 71
traffic psychology 80
traffic simulation 91, 141
traffic volume 28, 89, 141
traffic zone system 49
transit malls 107, 109, 136, 137
transport economics 80, 105
transport geography 80
transportation network companies
    166
transportation planning 61–63,
    73, 80, 81, 86, 102, 107, 113,
    152, 198
Transportation System
    Management 88
transport-based city planning 107,
    109
travel motivations 102
trip distribution 83
trip generation 83

uncertainty 107, 187, 188, 193,
    197
unmanned aerial vehicles 128
urban growth boundary 43, 70
urban redevelopment project 69
urban transport master plan 87

urban village 25
Urbanization Areas 43
Urbanization control area 43
Urbanization promotion area
  43
Utsunomiya 109, 110

vacation rentals 165
Variation Tree Analysis 96
virtual reality 143, 172
Vision Zero 98

walkable 115
walkable city 50
Willingness to Pay 81
Woonerf 28, 29

Yamanote Line 65, 66
Yufuin 118

Zone 30, 29
Zone of Possible Agreements 191
zoning 67, 70, 71

Printed in the United States
by Baker & Taylor Publisher Services